Neo-liberal Genetics

Neo-liberal Genetics:
The Myths and Moral Tales of Evolutionary Psychology

Susan McKinnon

PRICKLY PARADIGM PRESS
CHICAGO

Prickly Paradigm Press, LLC
5629 South University Avenue
Chicago, Il 60637

www.prickly-paradigm.com

ISBN: 0-9761475-2-1
LCCN: 2005908632

Second printing June 2009

Printed in the United States of America on acid-free
paper.

I. INTRODUCTION

In an era when the divisive politics of "family values" has created fault lines that threaten to render the United States asunder, evolutionary psychologists tell us that they have the single key to understanding the value of the family. At a time when ideas about sex and gender are rapidly changing and deeply contested around the globe, evolutionary psychologists tell a story about how gender difference was fixed forever in the depths of human evolutionary and genetic history. At a moment when the principles by which humans wish to organize society are up for grabs, evolutionary psychologists reduce social relations to a reflex of genetic self-maximization guided by the forces of natural selection. In a time when Anglo-American neo-liberal economics both dominates and is deeply resented and resisted in much of the world, evolutionary psychologists give us a theory of evolution that naturalizes neo-liberal values. In short, at a time when there is an urgent need for a nuanced understanding of the complexities and varieties of social life, evolutionary psychologists provide instead astonishingly reductive myths and moral tales.

Evolutionary psychology is one of those rare academic endeavors that has not only crossed disciplinary lines within academia but also broken the bounds of academia altogether to pervade the public media. Drawing from such fields as evolutionary biology, cognitive and experimental psychology, computational and game theories, and anthropology, it first developed as an academic inquiry, primarily inside departments of psychology. Its foremost advocates include, among others, John Tooby and Leda Cosmides, the co-directors of the Center for Evolutionary Psychology at the University of California, Santa Barbara; Martin Daly and Margo Wilson, who run a lab together in the Department of Psychology at McMaster University in Toronto; Steven Pinker, the Johnstone Family Professor of Psychology at Harvard University; David Buss, of the Department of Psychology at the University of Texas; and the journalist Robert Wright.

However, perhaps because the accounts of evolutionary psychology reflect familiar assumptions, the notion that our behavior is guided by psychological mechanisms that have deep evolutionary and genetic origins has quickly become part of the explanations in a host of different fields. Landscape architects talk about the deep structures of "prospect and refuge" that originated in the primal savannah environment and organize our appreciation of contemporary landscape design. The new field of evolutionary economics, David Wheeler tells us, is organized around the assumption "that much economic behavior may be the result of biologically based instincts to cooperate, trade, and bargain, and to punish cheaters." Kent Bailey and Helen Wood report a novel kind of psychotherapy, called evolutionary kinship therapy. This involves "first recognizing the various mismatch stresses the client is encountering (viz. disparities between the life circumstances of ancestral and modern humans), and then gently and compassionately helping the client first understand the problem and then make appropriate adjustments"—adjustments that bring current life circumstances back in line with presumed ancestral patterns. For legal sociobiologists, Herma Kay informs us, "biologically based behavioral sex differences can and should be used as the basis for legal distinctions supporting a conventional division of function by sex." And prominent jurists and legal scholars, such as Richard Posner, are using the assumptions of evolutionary psychology to think through issues relating to sex, gender, and family relations.

Despite the evident and wide-spread appeal of the ideas proposed by evolutionary psychologists, this pamphlet shows why, from an anthropological perspective, they are wrong about evolution, about psychology, and about culture. I make five basic arguments. I maintain that their theory of mind and culture cannot account for either the evolutionary origins and history or the contemporary variation and diversity of human social organization and behavior. More specifically, I demonstrate that assumptions about genetics and gender that underlie their theory of universal psychological mechanisms are not supported by empirical evidence from the anthropological record. I contend that not only their premises but also their evidence is so fundamentally flawed that their science is ultimately a complete fiction. I argue that this fiction has been created by the false assumption that their own cultural values are both natural in origin and universal in nature. And finally, I observe that this naturalization of the dominant values of one culture has the effect of marginalizing other cultural values and suppressing a wide range of past, present, and future human potentialities.

Contrasting Theories of Mind

What is most fundamentally at stake in the debates that surround evolutionary psychology is how we might think about the nature and processes of the human mind and culture. The theory of mind—and therefore the theory of culture—to which evolutionary psychologists subscribe contrasts sharply with the one to which most cultural anthropologists subscribe. The difference is not whether mental life is, in part, organically based or whether there is a complex developmental and interactive relation between the organism and the environment. Rather, as one commentator, Ted Benton, points out, what is at issue is "how much inherited 'architecture' there is in the human mind" and "whether sociocultural processes are understood as independent or reducible to inherited psychological mechanisms," which are, in turn, reducible to the principles of genetic maximization.

According to the theory of mind developed by evolutionary psychologists, the human mind operates through a multitude of psychological mechanisms that were developed in the Pleistocene environment of evolutionary adaptation. As answers to specific adaptive problems faced by our early ancestors, these mechanisms provide intricate content-laden instructions for particular forms of social behavior that are understood to be both innate and universal. While evolutionary psychologists acknowledge cultural diversity, they hold that the ultimate cause of human behavior and cultural formations is given by the adaptive logic of natural selection, which is driven by single-minded efforts to maximize individual reproductive success. If these innate mechanisms give shape to human behavior, then cultural patterns become the superficial dressing on an otherwise predetermined foundation. In evolutionary psychology, therefore, cultural ideas, beliefs, and values are epiphenomenal to, dependent upon, and reducible to the "real" genetic determinants of behavior. For evolutionary psychologists, the project is to delineate what Wilson and Daly call "the core mindset," which they argue underlies the manifest diversity of human culture.

By contrast, according to the theory of mind to which most cultural anthropologists subscribe, the mind is defined not by functionally specific mechanisms that solve particular adaptive problems but rather by general mechanisms that allow the brain to function as a flexible tool. It is these general mechanisms that enable humans to solve a wide range of problems in different contexts and to learn and create diverse cultural forms and behaviors. Indeed, anthropologists argue, the variable and fluctuating environments in which human evolution took place would have favored general mechanisms that allowed open programs of behavior and cognition, precisely the opposite of the function-specific modules posited by evolutionary psychology. As we shall see, research on human and other mammalian brains and from developmental psychology similarly supports the idea of the brain as a generalized learning and problem-solving device, one that allows for the creation of cultural worlds that are not reducible to a singular logic, and certainly not a logic of genetic proliferation. With such a theory of human mental capabilities, culture can be neither epiphenomenal nor reducible to genetic determinants or the logic of natural selection. Rather, it is the conceptual framework through which people make meaningful distinctions, understand the world, and act within and upon it.

The Calculus of Genetics and Gender

Because their theory of human behavior privileges an ultimate causality that relies upon genetics and natural selection, ideas about reproduction, sexuality, gender, marriage, and family are, necessarily, central to their accounts. Two assertions provide the framework for the stories they tell. First, evolutionary psychologists assert that kinship relations—indeed all social relations—follow from genetic calculations. That is, they follow from individuals' calculations of genetic proximity and of the utility of specific behaviors for the maximization of individual genetic endowments.

Within the more general genetic calculus of social relations, evolutionary psychologists posit a second, more specific, gendered calculus. Fundamental differences in the reproductive strategies of men and women, they argue, must follow from a biologically-based asymmetry in their relative parental investments. Thus it is assumed that men and women have different adaptive problems to solve as they seek to maximize their reproductive success. Given their relatively long-term reproductive investment, evolutionary psychologists contend, ancestral females confronted the problem of how to secure resources to support their offspring. By contrast, given their relatively short-term reproductive investment, males confronted the problem of how to access as many fertile females as possible.

Evolutionary psychologists posit a set of discrete, gender differentiated, and highly specific psychological mechanisms that, they assert, developed in response to such adaptive problems in the original environment of evolutionary adaptation. The exact nature of this environment is unspecified but is generally assumed to have been something akin to the

African savannah during the Pleistocene. Thus, men developed preference mechanisms for characteristics (such as youth, attractiveness, and shapeliness) that are supposedly cues to female reproductive value, while women developed preference mechanisms for characteristics (such as status, ambitiousness, and industriousness) that are cues to male resource potential. Although there is no evidence that such qualities were valued in the Pleistocene, these preference mechanisms are thought to have evolved through natural selection and to have elicited—through the operation of sexual selection—the evolution of the desired qualities in the opposite sex. The resulting preference mechanisms and gender qualities are assumed to constitute innate and genetically inherited psychological features that have not changed for millennia. In this way, evolutionary psychologists tell a particular story about the nature, origins, and universality of social categories—particularly those relating to sex, gender, family, and marriage—a story about the past that has consequences for how we think about the present and future possibilities of human relations.

The Science and Politics of Naturalization

Whatever claims might be made for the hard sciences, the human sciences are, by virtue of their subject, inevitably enmeshed in debates about the nature of social categories—and the disputes over evolutionary psychologists' ideas about the nature of sex, gender, and kinship are only the latest in a long series of such debates. On the one hand, the human sciences have a long history of naturalizing particular social categories and hierarchies—that is, of arguing that specific social categories and their associated hierarchies (gender, for instance, or race) are based in nature and are therefore inevitable and unchangeable. The measure of natural difference might be brain size or skeleton shape or skin color; it might be humors or hormones or intelligence quotients. More recently, the measure of natural difference has been genetic. The terms and the measures may have changed over time, but the process of naturalization has remained the same.

On the other hand, explorations in the human sciences have also had the effect of denaturalizing social categories and hierarchies—by revealing, for instance, that women's brains are neither controlled by their wombs nor limited in their capacities by their size; by discovering that race is a social, rather than a biological, category; by understanding that IQ measures social capital as much as, if not more than, any innate genetic endowment; by comprehending the difference between genetic potentiality and genetic determinism; by investigating the implications of social, political, and economic structures for the patterns of health, disease, reproduction, and death; and so on.

Anthropology has often participated in the process of naturalizing social categories and hierarchies. It has measured its share of skulls and brains; created its share of racial categories; crafted its share of evolutionary narratives that have sorted people into "savage" and "civilized." Anthropology has equally, if not, in my mind, more essentially and inevitably, been part of the process by which social categories and hierarchies have been denaturalized. In taking biological variation seriously, it has dismantled the biological categorizations of race. By taking linguistic variation seriously, it has demonstrated the symbolic foundation that underlies all human languages and thought. By taking seriously social categories such as kinship, sexuality, and gender, it has demonstrated their variability and their symbolic rather than merely biologic "nature." By tracking changes in the form and meaning of social relations and hierarchies both cross-culturally and over time, it has developed a heightened awareness of the changeability of human social arrangements. Such an awareness, properly understood, inevitably unties the knots of necessity that bind social relations into seemingly fixed forms. It argues that things could always be—and often are—otherwise.

In this pamphlet I consider a particular instance of this tension between naturalizing and denaturalizing discourses by focusing my attention on evolutionary psychology. On the one hand, I argue that evolutionary psychology is simply the most recent in a long line of reductive scientific narratives that have naturalized social categories and hierarchies—in particular, those of sex, gender, and kinship. On the other hand, I wish to bring these narratives into dialogue with the denaturalizing impetus of American cultural anthropology which has, since the time of Boas and his students, endeavored to recognize the integrity of alternative cultural understandings of the world and, in particular, of the relations of sex, gender, and kinship.

The Cold Hard Facts of Science

Evolutionary psychologists characterize themselves as a beleaguered minority. In language that resonates with that of the conservative right, they see themselves victimized by what Harvard Professor Steven Pinker calls an "establishment" of "elite" "intellectuals." Evolutionary psychology is the "real" science, seemingly the only real human science that is capable of dealing soberly with the obvious and cold hard facts of the human situation. Pinker contrasts evolutionary psychologists with their opponents, whom he caricatures as "radical scientists" who are "biased by politics" or "romantics" in the thrall of "feel-good moralism." Or, alternatively, they are deemed to be like religious fanatics whose views on the "holy trinity" are mere reiterations of "orthodoxy," "doctrine" and "mantras." Or, yet again, they are characterized as just plain lunatics, whose understandings of the world are simply "delusions," "madness," and "romantic nonsense."

I will make the argument that evolutionary psychology is bad science. But I contend that this is the case, not because I believe that "good science" is free of cultural content, while "bad science" is not. Rather, I will argue that this is the case because evolutionary psychologists have not been willing to put their fundamental premises and analytic categories at risk in an encounter with contrary evidence. They have, in fact, done the opposite. They have presupposed the natural universality of their own categories and understandings and, in so doing, they have effectively ignored and erased evidence that would have proved their theories wrong.

I therefore wish to explore the ways in which dominant cultural ideas and practices have been written into the infrastructure of evolutionary psychology. How has a self-consciously Victorian morality of sex, gender, and family relations been united with a neo-liberal economic ideology to transform the theory of evolution and natural selection into something I call neo-liberal genetics? I wish to investigate how these historically and culturally specific ideas are transformed, rhetorically, into cross-cultural and ahistorical universals. And I wish to examine how such culturally specific ideas, once naturalized into deep genetic and evolutionary history, have the effect of privileging and validating certain cultural ideas and social arrangements over others. And, finally, I wish to consider how this form of naturalization inevitably exerts—whether intentionally or not—a prescriptive and moral force.

The course of these explorations will follow several tracks. I analyze the rhetorical structures and strategies of the texts of evolutionary psychology in order to comprehend the presuppositions, analogies, and narrative structures that together accomplish the naturalization and universalization of a set of culturally specific understandings of the world. I evaluate the forms and quality of their evidence—including what they admit and dismiss as evidence, and the points where they misread evidence or substitute conjecture for evidence, and so on. And most importantly, I show that what they consider universal aspects of sex, gender, and family are actually dominant Euro-American conventions by contrasting these with the understandings of people in other cultures who think about and live these relations quite differently. One does not have to accept other people's truths as one's own in order to appreciate their effect in organizing the forms of social relations—including those that relate to differential reproductive fitness—in ways that challenge the "truths" of evolutionary psychologists. The difference between the positions of evolutionary psychologists and most cultural anthropologists is whether to accord other cultural ideas and practices an integrity and efficacy of their own making or whether to reduce them to a supposedly fundamental and universal logic that is, in the end, a reflection of historically specific Euro-American ideas and values.

II. MIND AND CULTURE

It is a central premise of classical evolutionary theory that natural selection *does not* involve intelligent, purposeful design. And it has been a central premise of anthropology that, since the evolved human brain is structured to allow flexible adaptations to multiple and changing environments, humans *do* have the capacity for intelligent, purposeful design—that is, they have the capacity to create, and transmit through learning, varied cultural orders, which shape the ways in which they understand, experience, and act in the world.

Evolutionary psychologists have reversed these premises: they grant active intellectual agency to genes and natural selection at the same time that they render humans as passive enactors of an agency that is not theirs and of a logic of which they need not be aware—indeed, of which they are deemed to be not conscious. Thus, the agency of genes and natural selection is presumed to manifest itself in the human mind as a set of innate evolved psychological mechanisms that provide detailed blueprints for specific units of gendered behavior that are always understood to have as their goal the maximization of genetic self-interest.

To the extent that human intellect and agency have been given over to natural selection and its genetic vehicles, culture is reduced to the superstructural effects of a more fundamental biological reality. This displacement is key to the reductive and fundamentalist character of evolutionary psychology. Therefore, in what follows, I analyze the rhetorical strategies through which evolutionary psychologists effect this transposition, consider how evolutionary psychologists characterize natural selection, genes, the human mind, and culture, and assess its consequences for a theory of mind and culture.

Natural Selection as Puppeteer, Policy Maker, and Programmer

Evolutionary psychologists grant natural selection the same qualities that have been attributed to god as the ultimate creator and source of universal order, design, truth, and purpose—a move that provides an ironic convergence with the ideas put forward by the contemporary proponents of "intelligent design." In the words of Robert Wright and Steven Pinker—perhaps the foremost popularizers of evolutionary psychology—natural selection is conceived of as the grand "puppeteer," the "ultimate policy maker," the "Blind Programmer," and the "engineer" that "designs" organisms, mental organs, and adaptations meant to maximize genetic proliferation. As the ultimate manager of genetic productivity, "natural selection does the thinking," has "goals" and "strategies," "works its will" and "executes its policies." Natural selection has both desires and the force to realize them: it "wants" and "gets" humans to do certain things—for instance, to be "nice to our siblings" but only "to *look like* we're being nice" to friends. Genes—the foot soldiers on the battlefield of natural selection—share in the creative agency of natural selection. Although evolutionary psychologists offer disclaimers that genes do not have agency, they nevertheless consistently portray genes as if they did. Genes are "selfish" and "mercenary" and, like natural selection, "have strategies" and "goals" (generally to propagate themselves). To that end, they compete, design, engineer, and build organisms and mental organs. They have perspective and "point of view"; they "want" and "get their way," "calculate," "control," "select," "speak" to us, and "counsel submission" and "spread themselves."

These metaphorical exuberances have two consequences. First, evolutionary psychologists have emptied the human mind of all the qualities—such as consciousness, agency, and creativity—that are normally connected with thinking. Second, they have transferred these qualities to genes and natural selection, despite the fact that it is self-evident, and a fundamental premise of evolutionary biology, that neither purposeful design nor intellectual agency is a quality of natural selection or genes. The consequence is a radical disjunction between the creative intellectual agency of the abstract forces of natural selection, on the one hand, and the mechanical, unthinking, realization of natural selection's "design" by humans who are rendered its vehicles, on the other. As Robert Wright so often says, "natural selection does the 'thinking'; we do the doing."

Natural Selection's "Shameless Ploy"

What then remains of human consciousness, agency, and creativity, if so much has been given over to genes and natural selection? As with other theories that posit an unconscious prime mover, in the accounts of evolutionary psychology humans end up as victims of false consciousness. Humans may have their own ways of explaining how and why they are loving, jealous, grieving, willful, deceitful, creative or destructive. But however people think about what they are doing and why, the "reality" always lies elsewhere. Evolutionary psychologists attribute both natural selection's invention of human emotions as well as the play of emotions in human life to a single ultimate, unconscious, and unvarying cause: the selfish logic of individual genetic proliferation. As Wright notes, emotions are "just evolution's executioners"—they are "proxies" for the deeper reality and "subterranean" logic of genetic calculation. Emotions are genetic inventions in the service of genetic proliferation. Pinker puts it this way: by "making us enjoy life, health, sex, friends, and children, the genes buy a lottery ticket for representation in the next generation with odds that were favorable in the environment in which we evolved. Our goals are the subgoals of the ultimate goal of the genes, replicating themselves." Evolutionary psychologists replace the content of our conscious goals, emotions, and understandings with what they claim is a deeper and truer reality that escapes everyone's consciousness except their own.

Not only is there a deeper truth underneath our conscious goals, emotions, and understandings, but the latter have evolved to hide the harsh, unyieldingly self-interested and calculating nature of that deeper truth. Natural selection ends up being seen as a devious and duplicitous force that has created a complex human consciousness capable of moral reasoning, pretensions to free will, love, generosity, and a range of other human emotions specifically in order to conceal the underlying "reality" of the mean and deeply amoral competition among genes to perpetuate themselves. While humans innocently assume that their lives are guided by moral principles, cultural understandings, and individual convictions, they are actually shaped, evolutionary psychologists tell us, by the cold calculation of genetic self-interest. From such a perspective, human moral codes are actually "a bit of genetically orchestrated sophistry," love and friendship are really "credit insurance," sympathy is "just highly nuanced investment advice," compassion is but another name for "our best bargain hunting," and good parenting is, after all is said and done, really shrewd portfolio management. With the deep cynicism characteristic of the accounts of evolutionary psychologists, Wright suggests that this ability to wrap crass self-interest in the garb of beneficent sociality and moral higher truths is natural selection's "shameless ploy." Culture, in such an account, is simply our means of self-deception in the service of genetic self-interest.

Mind as Mechanism and Module

What theory of mind, then, is appropriate to a narrative in which consciousness, agency, and creativity have been given to natural selection and genes, while humans have been made into the mechanical tools for a realization of an ultimate truth of which they are not aware? What theory of mind would be capable of translating an underlying agency of genetic competition into the forms of human sociality—the "thinking" of natural selection into the "doing" of human behavior? The mind appropriate to such a task is not possessed of general capabilities but rather of specific instructions engineered to address particular problems; it operates not consciously but unconsciously; and its operations are not culturally learned but genetically programmed.

Between the "reality" of genes' strategies to reproduce themselves and the actuality of human

behavior, evolutionary psychologists posit the existence of "mental organs," "mechanisms," and
"modules" that supposedly evolved in response to the
specific adaptive problems faced by our ancestors in
the Pleistocene environment of evolutionary adaptation. They postulate the development and existence of
such modules through a process they call "reverse
engineering." As Pinker explains, "[i]n forward-engi-
neering, one designs a machine to do something; in
reverse-engineering, one figures out what a machine
was designed to do." To reverse-engineer what they
see as a universal psychological mechanism, they look
for reasons why it would have solved a particular adaptive problem in the ancestral environment. For example, in order to explain the origins and existence of the
psychological mechanism that supposedly makes
women universally prefer men with resources, David
Buss and other evolutionary psychologists suggest that
"[b]ecause ancestral women faced the tremendous
burdens of internal fertilization, a nine-month gesta-
tion, and lactation, they *would have* benefited tremen-
dously by selecting mates who possessed resources.
These preferences helped our ancestral mothers solve
the adaptive problems of survival and reproduction."
To the extent that evolutionary psychologists can tell a
plausible story about the adaptive significance of a
proposed mechanism, it establishes for them its objective and universal reality. Thus, a hypothetical story of
origins is made to stand as proof of the existence and
universality of psychological mechanisms.

Such mechanisms and modules are thought to
have evolved not only through natural selection but

also through sexual selection, a process through which the mate choice of males and females effectively acts as a selective force in the development of secondary sex characteristics. Having developed in response to particular problems encountered in the original environment of evolutionary adaptation, moreover, these mechanisms are assumed to constitute innate and genetically inherited psychological features that have not changed since the Pleistocene.

Evolutionary psychologists reject the idea that the evolved human brain manifests a generalized capacity to create a wide array of cultural forms and learn a diverse range of behaviors. Just as our physical organs are genetically programmed and function without learning, so too do our mental organs. "We do not learn to have a pancreas," Pinker astutely observes; and he immediately goes on to assert, "and we do not learn to have a visual system, language acquisition, common sense, or feelings of love, friendship, and fairness."

Evolutionary psychologists speculate that a series of psychological mechanisms developed that were especially designed to solve the particular "adaptive problems" of our Pleistocene male and female ancestors. These mechanisms, or modules, are gendered, functionally specific, and provide detailed blueprints for psychological preferences and behavior. For instance, evolutionary psychologists argue that, since one of men's central adaptive problems is to determine which females are fertile, they have developed preference mechanism for those characteristics—such as youth, attractiveness, and shapeliness—that supposedly are reliable signs of female fertility.

Similarly, since one of women's central adaptive problems is to find men with resources, they have developed preference mechanisms for those characteristics—such as status, ambitiousness, and industriousness—that are supposedly reliable signs of resource potential and solve the problem of assessing which males are suitable partners capable of investing in their offspring. Such mechanisms are conceptualized as "dispositions, decision-rules, structures, processes" that reside "within the organism," that process information in accordance with a cost-benefit utilitarian logic focused solely on reproductive success, and that "produce output" in the form of behavior that is an appropriate response to adaptive problems related to genetic maximization. These mental organs, in short, do not enable humans to respond flexibly or creatively to situations. They act, rather, as mechanical translators of the "thinking" that natural selection does into the "doing" that humans do—without requiring that humans actually think about anything at all.

Given the specificity of the mental organs hypothesized by evolutionary psychologists—as well as their assumption that it is natural selection that is doing most of the "thinking"—it is not at all surprising that the metaphors of mind they employ evoke the precision of machines and computers. Machines and "mechanisms" are designed to endlessly replicate the same procedure and produce the same output; computers are programmed with an underlying code; and modules are self-contained, standardized components of an overall structure. Indeed, the mental organs of humans have, according to evolutionary psychologists, been using the same mental mechanisms to produce the same behavioral "output" for millennia.

The stability of the specific psychological modules posited by evolutionary psychologists is also evident in their common characterization of the brain as an architectural edifice. Indeed, to the question, "how much inherited 'architecture' is there in the human mind?", evolutionary psychologists answer not only with a fully built structure but also with a structure whose interior is furnished in exquisite detail. Yet as Steven Rose points out, the architectural metaphor "which implies [a] static structure, built to blueprints and thereafter stable, could not be a more inappropriate way to view the fluid dynamic processes whereby our minds/brain develop and create order out of the blooming buzzing confusion of the world which confronts us moment by moment."

Indeed, since evolutionary psychologists posit genetic self-maximization as the fixed "subterranean" prime mover of human behavior, they must furnish the human mind with specific structures and furnishings that will express only this particular motivation and no other. The hypothesis that specific content-laden psychological modules exist is a requirement of a theory that has made the human mind a passive enactor of a singular and unconscious truth and of an abstract agency. Indeed, what is required is an "architecture" or "machine" that is capable of erasing the multiplicity of human truths, realities, and motivations and transforming them into a singular truth, reality, and motivation. As we shall see, this is precisely what the "mental modules" and "mechanisms" accomplish.

Fixed Architecture vs. Neural Plasticity

I pause here to make clear that other models differ significantly from this portrait of the brain as an unchanging architectural edifice constructed out of fixed, functionally specific modules. Kathleen Gibson, a biological anthropologist who specializes in the evolution and development of primate and human brains and cognition, has recently summarized the arguments against the model of the brain put forward by evolutionary psychologists. She focuses on several points that are relevant to our inquiry: genetic economy; neuroanatomical generalization; epigenesis; and mental constructional processes.

The well-established fact that most genes are pleiotropic—that is, they have far more than one phenotypic effect—provides a strong argument against a one-to-one correspondence between a gene and a behavioral trait. Thus any complex human behavior will be influenced by multiple genes, and a given gene will contribute to the realization of multiple complex behaviors. Moreover, Gibson points out that the human genome is comprised of "approximately 30,000 genes and that humans and chimpanzees differ in only 1.6 percent of their DNA." There are, quite simply, not enough genes in the human genome to account for both the physical and behavioral differences between humans and chimpanzees and for the large number of functionally discrete modules posited by evolutionary psychologists. "What is needed," Gibson argues, "is not a theory that one gene = one mental module = one complex behavior but a theory of how a small number of genes can construct a complex brain and enable a diversity of behaviors."

While scientists have known for some time that specific regions of the brain are associated with certain behaviors and cognitive capacities, their understanding of how those regions shape behavior has become more nuanced. Indeed, recent research on brain neuroanatomy suggests that what is localized in different regions of the brain are "mechanisms that may contribute to varied behavioral and cognitive domains" rather than to specific ones. For instance, "Broca's area was once thought to control speech, then thought to control syntax, and later postulated to function in the hierarchical organization of both speech and manual behaviors...." Such reformulations of our ideas about how the areas of the brain work direct us away from function-specific psychological modules and toward more general processing mechanisms.

The fixed module model of the brain advocated by evolutionary psychologists is further challenged by epigenesis—the interaction of genes with environmental input—and the well-documented neural plasticity of the brain. Gibson notes that "maturing mammalian and human brains are classic epigenetic systems that acquire species-typical neural processing mechanisms and behavioral capacities through interactions between genes and environmental inputs." Evidence for epigenesis and neural plasticity includes, for instance, the hyper-development of one sensory-motor capability (for instance, aural, visual, or tactile) in the absence of or assault on another; the shift of linguistic and motor functions from one hemisphere to the other under the stress of childhood brain damage; the effects of nutrition and intellectual stimulation on brain development and function; and the "massive overproduction of neurons and synapses that are then fine-tuned by environmental input." The ability of environmental factors to shape the expression of genes over time is evident in recent research on identical twins that was carried out by Mario F. Fraga and others at the Spanish National Cancer Center in Madrid. This research, which focused on the DNA of more than 40 pairs of identical twins, demonstrates that the epigenetic profile of identical twins varies both over a lifetime and across social environments. *The Washington Post* science writer, Rick Weiss, summed up the results of the research:

> They found that young twins had almost iden-tical epigenetic profiles but that with age their profiles became more and more divergent. In a finding that scientists said was particularly groundbreaking, the epigenetic profiles of twins who had been raised apart or had especially differ-ent life experiences—including nutritional habits, history of illness, physical activity, and use of tobacco, alcohol and drugs—differed more than those who had lived together longer or shared similar environments and experiences.

The fact that the human brain and all complex phenotypic traits develop through epigenesis makes it difficult to disentangle the effects of genes and environment in a developed adult brain. Although evolutionary psychologists do not appear to be interested in actually locating specific genes or areas in the brain that correspond to their proposed modules and mechanisms, the significance of epigenesis and neural plasticity for their theory is clear. Gibson observes that, "[e]ven if at some future date adult brains were found to contain [function-specific] regions dedicated to cheater detection or reciprocal altruism, that would be insufficient evidence to draw sweeping conclusions about their genetic or developmental determination."

In any case, it is important to remember that epigenesis and neural plasticity, as such, can not differentiate human mental capacities from other mammals, including our closest relatives, the great apes. Rather, Gibson argues, it is the ability of the human brain to process significantly larger amounts of information and to do so in hierarchically ordered structures—what she calls "mental construction"—that differentiates humans from the great apes. "Specifically," her research shows, "the increased information-processing capacity of the human brain allows humans to combine and recombine greater numbers of actions, perceptions, and concepts together to create higher-order conceptual or behavioral constructs than do apes." This ability to embed multiple concepts, perceptions, or actions into hierarchically ordered constructs is what makes it possible to carry out a wide range of technical, linguistic, and social tasks—from building complex tools and structures, to constructing sentences, to inferring others' intentions (for instance, to tell the truth or deceive, to joke, or to be ironic or sarcastic). The fact that this capacity develops during maturation accounts for the differential abilities of human infants and adults; its full development in human adults differentiates them from the great apes.

There are several reasons why an appeal to a more general, underlying process of mental construction is a more satisfying account of the human mental capacities than one that relies upon a host of fixed and functionally specific mental modules. Gibson sums it up this way:

> A mental constructional model... accords with the principles of pleiotropy, genetic economy, and epigenesis, and it explains our abilities to devise creative solutions to novel problems. In contrast, models that postulate genetically controlled, functionally dedicated neural modules for each problem encountered during our evolutionary history are genetically expensive, violate principles of pleiotropy, and cannot explain why we can solve problems and adapt to environments not encountered by our ancestors.

The model of the brain outlined by Gibson makes sense of the anthropological and archaeological evidence of human creativity that is manifest in the diversity of cultural ideas, beliefs, and practices across the globe, the variability of individual understandings and behavior, and the historical transformation of cultures over time.

Shady Accounting Genes

Although the evidence just cited suggests that even generalized mechanisms in the human brain cannot be accounted for by reference to genetics alone, evolutionary psychologists argue that specific mental mechanisms—each of which solves a particular problem in a delimited domain—are innate. As Donald Symons states: "One can, hence, properly say that mechanisms specialized *for* mate preferences came to exist in human psyches and that genes *for* mate preferences came to exist in human gene pools." Others draw an analogy between "innate psychology" (as opposed to a "manifest psychology") and the human genotype (as opposed to the human phenotype), implying that psychological mechanisms are not only "like" but also actually part of the human genotype. Indeed, Tooby and Cosmides assert that "[c]omplex adaptations are intricate machines that require complex 'blueprints' at the genetic level."

Most evolutionary psychologists clearly state that there is not a one-to-one correlation between any specific gene and any particular mental organ, module, or mechanism. Nevertheless, their narratives are replete with conjectures that involve named genes and modules that correlate with remarkably specific behaviors—a narrative strategy that is misleading at best. As one reads along, one forgets the caveat and hears only the repetition of one-to-one specificity of the genetic-behavioral link.

Evolutionary psychologists do not shy away from being quite specific about all the behavioral traits that are deemed to be heritable. Pinker argues that not only general talents and temperaments ("how proficient with language you are, how religious, how liberal, or conservative"), but also personality types ("openness to experience, conscientiousness, extroversion-introversion, antagonism-agreeableness, and neuroticism") as well as highly specific behavioral traits ("such as dependence on nicotine or alcohol, number of hours of television watched, and the likelihood of divorcing") are all inherited—i.e., are genetically encoded. All of this notwithstanding the notable absence of nicotine, alcohol, television, and perhaps even marriage, let alone divorce, in the Pleistocene social scene!

Moreover, evolutionary psychologists do not hesitate to take a hypothesis or counterhypothesis about the nature and evolution of human behavior and reify it in the form of a named gene or module, which has the rhetorical effect of transforming the *fiction* of a genetic correlation into a "scientific" *reality*. Among the many named genes that are featured in the texts of evolutionary psychology one finds:

- a "fidelity gene"
- an "altruism gene"
- a "gene that leads a chimpanzee to give two ounces of meat to a sibling"
- a "gene counseling apes to love other apes that suckled at their mother's breast"

- a "gene that repaid kindness with kindness"
- a "club-forming gene"
- a "gene that made a child murder his newborn sister"
- a "gene that made a fifteen-year-old want to nurse"
- a "gene predisposing a male to be cuckolded"
- a "gene that allowed a male to impregnate all the females"
- a "gene for joining in the game."
- "genes counselling submission"
- genes that instill drives such as ambition, competitiveness or feelings such as shame or pride
- "shady accounting genes"
- "reciprocal-altruism genes"
- "genes that direct altruism toward altruists"
- "genes for helping relatives"
- "genes for resisting... roles"

Among the many named modules featured in the texts of evolutionary psychology one finds:

- a "love of offspring module"
- an "attracted to muscle module"
- an "attracted to status module"
- an "age-detection system"
- a "mate-ejection module"
- a "cheater-detection module"
- a "mate-killing module"
- an "evolved homicide module"

There are several problems with this form of genetic reification of specific human behavioral traits. First, and not least of all, no genes or modules for specific human behavioral traits, temperaments, or personality types have been found. Second, this empirical void does not appear to deter such rhetorical reifications and fabrications. As Steven Rose points out, the "spread of such theoretical pseudo-genes, and their putative effect on inclusive fitness can then be modeled satisfactorily *as if* they existed, without the need to touch empirical biological ground at all." Indeed, anthropologists Stefan Helmreich and Heather Paxson note that evolutionary psychologists repeatedly "offer hypotheses and later refer to them as if they had been proven." Third, the Human Genome Project's recent discovery that the human genome has far fewer genes than formerly imagined—some 30,000—militates against the supposition that a host of functionally specific psychological mechanisms could possibly have discrete genetic correlates. And finally, such representations fundamentally misrepresent the nature of genes and their relation to evolution. The evolutionary geneticist, Gabriel Dover, who specializes in molecular evolutionary processes, observes:

> Genes are not self-replicating entities; they are not eternal; they are not units of selection; they are not units of function; and they are not units of instruction. They are modular in construction and history; invariably redundant; each involved in a multitude of functions; and misbehave in a bizarre range of ways. They coevolve intimately and interactively with each other through their protein and RNA products. They have no meaning outside of their interactions, with regard to any adaptive feature of an individual: There are no one-to-one links

between genes and complex traits. Genes are the units of inheritance but not the units of evolution: I shall argue that there are no "units" of evolution as such because all units are constantly changing. They are intimately involved with the evolution of biological functions, but evolution is not about the natural selection of "selfish"genes.

Whether or not evolutionary psychologists actually believe there are one-to-one correlates between genes and complex human behavioral traits, their reification of such traits in named genes naturalizes the results of conjecture, cloaks them in the authority of scientific truth, and makes it impossible to consider alternative origins and modes of transmission for the complex array of human behavioral traits.

The Rationality of Absolutely Everything

Evolutionary psychologists pride themselves on the parsimony of their explanations—which are, indeed, parsimonious in the sense that they appeal to the same underlying genetic rationality to explain everything. Even the most irrational and destructive acts can be made to seem rational and productive. Take the example of jealousy. In his book, *The Dangerous Passion: Why Jealousy is as Necessary as Love and Sex*, David Buss argues that this seemingly irrational emotion is attributed primarily to men and is seen as a rational adaptive response to men's perpetual uncertainty about paternity. Evolutionary psychologists argue that jealousy manifests itself in a man in response to the possibility of a "real threat"—i.e., that his partner may be having sex with someone else and, consequently, that he may not only lose an opportunity to reproduce with his partner (or find another) but also potentially "waste" his paternal investment on children not genetically his own.

Buss conceptualizes jealousy as a kind of unconscious smoke detector, although this is not his metaphor. Jealousy (not the man) detects the smoke of infidelity and sets off the alarm when there is a fire of infidelity (and often, as in false alarms, even when there is not). The infidelity detector is run in accordance with an unconscious reproductive calculus, and therefore the man need not be aware of the "cues" or understand why he is feeling jealous. Jealousy sounds the alarm in order to elicit appropriate actions from a man's partner that will protect him from (further) cuckoldry and the potential waste of his reproductive potential and investments. So, the "rationality" of jealousy is supposedly in its function as a detection mechanism that "sees" the true state of things (where the vision of the man might be clouded) and effects a recalibration of the relation to protect his (but not necessarily the woman's) reproductive assets.

The hypothesis of a "subterranean" genetic logic allows Buss and other evolutionary psychologists to see even those actions that seem most irrational— for instance, where jealousy persists as an obsession and involves violence despite the fidelity of the partner—as fundamentally rational and adaptive. Buss is able to draw this conclusion even from evidence that contradicts it. He reports that

[o]ne study interviewed a sample of battered women and divided them into two groups: one group had been both raped and beaten by their husbands and one group had been beaten but not raped. These two groups were then compared with a control group of non-victimized women. The women were asked whether they had "ever had sex" with a man other than their husband while living with their husband. Ten percent of the non-victimized women reported having had an affair; 23 percent of the battered women reported having had an affair; and 47 percent of those who were battered and raped confessed to committing adultery. These statistics reveal that infidelity by a woman predicts battering behavior in men.

The fact that violent jealousy appears to operate even when a *majority* of women—77 percent of battered women and 53 percent of battered and raped women—*had not committed adultery*, does not seem to register. Without skipping a beat, Buss goes on to argue that violent jealousy is "rational" because when "men maintain a credible threat,... [they lower] the odds that their partner will commit infidelity or defect from the relationship" and thereby cause the men to lose the battle for differential genetic proliferation. Going further, Buss and Duntly postulate the evolution in men of a "mate-killing module," which, following a cost-benefit analysis, would dictate the killing of a "mate" for various reasons: to staunch the costs of paternity uncertainty and the misdirection of parental investments (i.e., when a man presumes his spouse to be pregnant with another man's child), or to avert a mate's permanent defection to another man. Homicide is thus seen simply as an "adaptive method of reducing [the reproductive] cost" to the murderer.

Given the supposition that there is one true reality underlying all human behavior and that a particular logic of rational choice is the ultimate cause of all human behavior, there can, by definition, be no human behavior—no matter how destructive—that cannot be rendered logical and rational. One need only posit the existence of a suitable "module" to effect the translation between genetic utility and behavioral manifestation. The fact that completely contrary behaviors—caring for and killing one's spouse and child—are accounted for by the same fundamental logic means, as Hilary Rose notes in another context, that "selection explains everything and therefore nothing."

But this "parsimony" of explanation comes with a triple paradox: it creates choices that are not choices; it posits individuals that are not individuals; and it imagines cultural effects that are independent of culture.

The Choice That Is Not a Choice

"Choice" is an important value in Euro-American cultures, one that generally signifies the realm of culture rather than nature. It is a paradox that, in a theory so thoroughly grounded in a neo-liberal ideology of "rational choice" and individual self-interest, individuals end up being denied the ability and agency to make conscious choices. In the narratives of social life written by evolutionary psychologists there is a lot of "thinking" going on—decisions are made, problems are solved, rules are followed, costs and benefits are analyzed, choices are made, and goals are met. Yet humans are not doing the "thinking" or "choosing." Evolutionary psychologists constantly tell us that humans do things without being conscious of the subterranean genetic logic that motivates their actions. Humans think, make decisions and choices, solve problems, and analyze costs and benefits in the same way as their sweat glands control thermal regulation—without needing to be conscious of the process. "Indeed," Buss notes, "just as a piano player's sudden awareness of her hands may impede performance, most human sexual strategies are best carried out without the awareness of the actor." If sweat glands are the model for how humans make choices, it's hard to know why brains evolved in the first place. We could have just used our ovaries and testes to "think" about the world.

While such a proposition seems laughable, it is precisely how evolutionary psychologists imagine that we do our thinking. The active agents in their narratives, the agents that do the calculations, weigh the costs and benefits, and make the choices, are rarely conscious persons and are quite often microscopic physiological entities—such as sperm, eggs, hormones, or genes.

One of evolutionary psychology's central tropes—that males compete and females choose—is thought to play out most significantly at the microscopic level. Thus, the male ejaculate is seen as an army or "horde" that does "subterranean battle." It is made up of various specialized agents with distinct missions: the mission of the "egg-getter" sperm is to swim fast and reach the goal as quickly as possible, while the mission of the "kamikaze" sperm is to take out the enemy and sacrifice themselves to the greater good of their quicker comrades-in-arms.

Whereas male sperm "advertise" their qualities, females (or, presumably, their eggs) unconsciously manipulate sperm, "sense" which is a good match, and pick and choose the best among them. This has been termed "cryptic female choice"—cryptic, indeed, since even the woman herself is not conscious of the choice. Interestingly, the mechanism by which a woman unconsciously "chooses," or "controls" which partner's sperm she will "accept" or "reject" turns out to be, according to Devendra Singh and his (all male) colleagues, the frequency and timing of orgasm. But since, by their own reckoning, women experience orgasm only 40% of the time and, one should note, "timing" is hardly a precise science in any case, women's unconscious choice turns out to be not much of a choice at all. It does not matter, however, because the woman did not know she was choosing: her eggs and hormones had conscious agency.

In their willingness to subsume human agency to the singular logic of genetic calculation, evolutionary psychologists have erased choice as a function of human consciousness and reinvented it as a function of human tissues, physiological processes, and the more general process of natural selection. Although

the tremendous cultural salience of economic "rational choice" has been retained in the rhetoric of evolutionary psychology, it has been divorced from human
consciousness and rendered the property of the
"thinking" that natural selection does for us. This
invisible hand of natural selection is thought to operate in biology much as the "invisible hand of the
market" is presumed to operate in the economy.

The Individual That Is Not an Individual

If choice turns out not to be much of a choice, the
individual turns out not to be much of an individual,
either. Because there is only one prime mover and
only one logic that counts in the explanations of
evolutionary psychologists, they do not need a theory
of individual motivation and they do not need to
appeal to the details of individual life histories—both
of which are deemed irrelevant to the ultimate explanations favored by evolutionary psychologists.

Humans are obviously capable of a range of
different emotional states, lines of reasoning, and forms
of creativity. They are capable of mild or morbid jealousy or no jealousy at all, of peaceful mediation,
unspeakable violence, or outright murder. Evolutionary
psychologists are incapable of explaining at least three
aspects of the differential response of individuals to a
particular situation—for instance, the presumed infidelity of one's sexual partner or spouse. First, evolutionary psychologists are not able to explain why individuals deviate in the expression of a particular cultural

norm. In cultures where extramarital sexuality is openly permitted and the expression of jealousy not encouraged, why do some individuals manifest jealousy nonetheless? And, conversely, why, in places where extramarital sexuality is strongly permitted and the expression of jealousy openly sanctioned, do some people not express jealousy at all? Second, they cannot explain why there are so many possible responses to the same act—extramarital sexuality of one's spouse. Why does it get framed as infidelity for some and sexual freedom for others? Why is cold withdrawal a sufficient expression of jealousy for some men and murder hardly suffices for others? And third, evolutionary psychologists can not explain the manifest inappropriateness of the response in some cases—i.e., the apparent discrepancy between cause and effect. For example, even if one buys the argument that men's battery of women is an adaptive response to women's infidelity, they are incapable of explaining the huge gap (77%) between the number of cases of battery and the number of cases that actually involve infidelity on the part of women.

An adequate theory of sexual jealousy ought to be able to explain why jealousy is or is not expressed; why, when it is expressed, it takes significantly different forms; and why it is expressed when there is no objective cause. Evolutionary psychologists can not explain these things because there are clearly other logics at work here that escape the particular "rationality" of reproductive advantage—logics that, if we are to comprehend them, require attention to the specificity of individual histories and motivations as well as cultural values and understandings.

The Culture That Is Not Culture

Evolutionary psychologists skirt the evidence of cultural variation by casting it as a kind of "manifest" or surface structure that is subordinate to an innate or deep structure of genetically determined psychological mechanisms—what John Tooby and Leda Cosmides describe as the difference between the "phenotype" and the "genotype." While acknowledging cultural complexity and diversity, Margo Wilson and Martin Daly argue for "the ubiquity of a core mindset, whose operation can be discerned from numerous phenomena which are culturally diverse in their details but monotonously alike in the abstract."

The distinction between deep structure and surface structure—genotype and phenotype—has the added attraction that it can be made to account for absolutely anything. When the "genotype" preference is not expressed, this can be explained by environmental (read cultural) forces that trigger an alternate "phenotypical" form. In David Buss' survey of 37 cultures, he notes that, contrary to prediction, Zulu men favor ambition and industriousness in women more than women do in men. In this case, it is not the supposedly innate mate preferences but rather the culturally specific division of labor in Zululand that is appealed to in order to account for the reversal in prediction. It is therefore only by resorting to heterogeneous arguments in an ad hoc manner that the account can be supported: innate structures account for women's preference for male ambition and industriousness (read resources) in some cultures, but cultural structures account for men's preference for female ambition and industriousness in others.

Evolutionary psychologists are forced into such ad hoc arguments concerning the origin of cultural variation for several reasons. They have rejected a theory of mind that posits generalized mental capabilities in favor of one that stipulates functionally specific, content-laden psychological mechanisms. They have, no matter what they say about nature and nurture, dissolved the dynamics of culture into biology and the utilities of genetic maximization. And they have elevated their own, culturally and historically specific understandings about the world into cross-cultural universals that, they are surprised to find, do not always accord with the manifestations of other cultures.

III. INDIVIDUAL AND SOCIETY

Evolutionary psychologists build their theories of psychological preferences upon a particular form of reductionism—I call it genetic individualism—that was first formulated as part of sociobiology in the 1970s. By genetic individualism, I mean a conception of human social life that reduces social relations and human behavior to the product of self-interested competition between individuals. These individuals (or their genes) calculate their interests according to a cost-benefit logic that has, as its goal, the proliferation of genetic endowments through natural selection. The notion of genetic individualism relies, explicitly or implicitly, upon the cultural values of neo-liberal economic theory: that social relations can be reduced to market relations; that the "public good" should be replaced by individual responsibility and social services privatized; that profit and capital should be maximized through the deregulation of markets—that is, that competition should run its course unchecked—in a "race to the bottom"—regardless of the social consequences.

Starting from a position of radical genetic individualism, one of the central conundrums that evolutionary psychologists must address—like sociobiologists before them—is the existence of social life, itself, particularly those forms of social behavior that are not evidently self-interested. What, then, are the building blocks of genetic individualism and how are they rhetorically assembled? How do evolutionary psychologists conceive of the generation of social relations? And, most importantly, to what extent is it actually possible to account for the varieties of human social relations by reference to either the genetics or the individualism of genetic individualism? Does the genetic logic of evolutionary psychology stand or fall when it is made to confront the empirical evidence of human cultural diversity?

Genetic Individualism and the Problem of the "Social"

For evolutionary psychologists, as much as for sociobiologists, life is about the workings of natural selection, and natural selection is about the differential proliferation of genes among individuals from one generation to the next. Individual genetic self-maximization is the primary motivation of behavior, and competition is the primary form of social relation as the single-minded (if unconscious) goal of all individuals is to maximize the proliferation of their own genes in the subsequent generations.

To the extent that no two people (except for identical twins) have the same genetic makeup, evolutionary psychologists argue, a person's genetic interests always inherently conflict with those of others. Since they see this conflict as the primary fact of human existence, it is inevitable that they envision social life as essentially a form of warfare and a series of battles. Thus natural selection is seen as a "battleground," populated with various actors who "pose threats" and defend themselves with distinctive "arsenals" of cognitive and emotional "weapons." While it is no surprise that males are imagined to "do battle" with males, and females with females, evolutionary psychologists picture a much more extensive battlefield. Because they see the genetic interests of males and females to be essentially at odds with one another, they imagine that much of human behavior is shaped by a "battle between the sexes," as males and females engage in a perpetual evolutionary "arms race" of opposing reproductive strategies. However, the battle is waged not only between the sexes but also even between parents and children (a battle that is thought to begin in the mother's womb) and between siblings. Thus, beginning with the presupposition that human life is fundamentally a Hobbesian war of each against all, the problem then is how to conceptualize forms of sociality that are not obviously self-interested and competitive.

From a foundation in the genetic interests of the individual, evolutionary psychologists construct a primary order of social relations—one that follows from a natural rather than a cultural logic. Sexual reproduction is understood to result in a natural calculus of genetic relatedness or kinship proximity. For instance, one shares more genes with, and is therefore closer to, one's biological siblings than one's siblings' biological children. One can thereby plot degrees of genetic closeness and distance as one moves out from any individual to more and more distant relations.

Following sociobiologists, evolutionary psychologists make the critical supposition that social relations follow directly from genetic relations. They posit a one-to-one relationship between genetic and social relationships that makes kinship categories clear-cut and self-evident. As Pinker quips, kinship "relationships are digital. You're either someone's mother or you aren't." But it is not only the *categories* of kinship relation but also the *behaviors* appropriate to various categories of kin—for instance, love, nurturance, altruism, solidarity—that are assumed to follow directly and without mediation from the degree of genetic relatedness: the higher the degree of genetic relatedness, the higher the degree of such behavior. Evolutionary psychologists argue that, because people are driven to maximize their own reproductive success, they wish to "invest" first and foremost in their own genetic children and do not want to "waste" resources on children that are not genetically their own.

Evolutionary psychologists expand this radically exclusive and restrictive vision of human sociality by reference to the principles of kin selection and inclusive fitness, which make it possible to include more distant relatives in the social universe. Inclusive fitness stipulates that, because biological relatives share genes, altruistic behavior toward relatives contributes to the proliferation of an individual's genes by promoting the viability of their shared genes. The mechanism of kin selection specifies that the degree to which altruistic behavior contributes to an individual's fitness is proportional to the percentage of genes shared with those who benefit from the altruistic act. Yet, even given these principles, the social relations that evolutionary psychologists are able to conceptualize are restricted to those who have some degree of genetic relation.

Evolutionary psychologists therefore refer to a secondary order of social relations—one they see as fundamentally "unnatural" and contrived—to connect those who lack any genetic relation whatsoever. "The love of kin comes naturally," Pinker asserts, while "the love of non-kin does not." To generate a more encompassing order of sociality, evolutionary psychologists rely upon the concept of reciprocal altruism—a system in which an individual performs an altruistic act for another with the expectation of future reciprocation, a system that, it is understood, requires a mechanism for the detection of cheaters. This allows individuals to cooperate or exchange goods or services with the understanding that each will benefit from the relation over time. But even here, this expanded realm of sociality is understood to be based on individual self-interest and a cost-benefit analysis that ultimately refers back to the natural logic of genetic proliferation.

In his 1976 book, *The Use and Abuse of Biology: An Anthropological Critique of Sociobiology*, anthropologist Marshall Sahlins points out that reciprocal altruism poses an unacknowledged paradox for sociobiology, and the same point can be made in reference to evolutionary psychology. Commenting upon Robert Trivers' 1971 account of reciprocal altruism, Sahlins observes that:

> Trivers becomes so interested in the fact that in helping others one helps himself, he forgets that in so doing one also benefits genetic competitors as much as oneself, so that in all moves that generalize a reciprocal balance, no *differential* (let alone optimal) advantage accrues to this so-called adaptive activity.... Hence the apparent nonfalsifiability of the argument: both altruism and nonaltruism are gainful, thus "adaptive"—so long as one does not inquire further whether the gain is also relative to other organisms.

The paradox reveals the contradictions inherent in attempting to account for the forms and dynamics of social life from a starting point of individual genetic self-interest.

Yet the narratives of evolutionary psychology seek to illuminate precisely this problem: how to generate the possibility of a more expansive sociality from what Robert Wright calls the original "meanness [that supposedly] pervaded our evolutionary lineage."

Beginning from the assumed cost-benefit calculus of individual genetic self-interest, the "natural" logic of genetic relatedness is presumed to translate directly into kinship categories and correlated kinship behavior. Once kin selection and inclusive fitness made evident the benefits of altruism among genetically related individuals, evolutionary psychologists argue, it was possible to generalize a self-interested altruism to genetically unrelated individuals in the form of reciprocal altruism.

It is significant, moreover, that evolutionary psychologists understand these forms of social relation to be transmitted not socially but genetically. In a remarkable passage in which he tracks the movement from the original "meanness," noted above, to the development of altruism and reciprocal altruism, Robert Wright posits a plethora of genes along the way. Thus he maps the trajectory by way of "a gene counselling apes to love other apes that suckled at their mother's breast" (hardly a sign of genetic siblingship, one should note, since women can and do suckle children that are not genetically related to them!), "genes directing altruism toward sucklers," and "genes directing altruism toward altruists." Ultimately he imagines that a "gene that repaid kindness with kindness could thus have spread through the extended family, and, by interbreeding, to other families, where it would thrive on the same logic." In presupposing the causal primacy of genes and rejecting the centrality of learning and creativity in human life, Wright is thus forced to posit this bizarre proliferation of fantasy genes to fabricate an account of the origins of human sociality.

In the end, there is a double genetic determinism animating the accounts of evolutionary psychologists. Not only do they imagine that the specific forms of human social relations develop *in response to* the motivations of genetic self-maximization but they also assert that these forms of social relation *are actually coded in* the human genome and spread from family to family through interbreeding. The effect is the erasure of culture, since both the creation and the transmittal of cultural understandings and behavior are presumed to be genetically orchestrated.

The Poverty of the Genetic Calculus

The question, then, is to what extent can such a theory—one that rests on a calculus of genetic self-interest, an ideology of individualism, and the erasure of culture as an independent level of analysis—actually account for the diverse modalities of human kinship and social relations? It turns out that one does not have to probe the empirical evidence far to comprehend that kinship is everywhere mediated by cultural understandings and can never be the simple precipitate of a universal "natural" genetic calculus.

In *The Use and Abuse of Biology*, Sahlins long ago demonstrated that the logic of kinship groupings—*whatever their form*—always and everywhere challenges the logic of genetic self-maximization and kin selection. Even if we begin from a descent-based ideology that presupposes a genealogical grid, it is impossible to generate the range of human kinship

groupings from a calculation of genetic proximity.
Take, for instance, a common means of delimiting kin
groups through the principle of unilineal descent—in
which descent is traced either through the male line to
constitute patrilineal groups or through the female
line to constitute matrilineal groups. Either way, given
a rule of exogamy, which requires marriage outside of
the lineage, some close genetic kin will end up in
other groups while genetic strangers and more distant
genetic kin will end up in one's own group. As Sahlins
notes,

> Over time, the members of the descent unit
> comprise a smaller and smaller fraction of the
> ancestor's total number of genealogical descendants,
> diminishing by a factor of 1/2 each generation.
> Assuming patrilineality, for example, and an equal
> number of male and female births, half the
> members in each generation are lost to the lineage,
> since the children of women will be members of
> their husband's lineage.... by the third generation
> the group consists of only 1/4 the ancestor's genea-
> logical kin, by the fifth generation, only 1/16, and
> so on. And whereas those of the fifth generation in
> the paternal line may have a coefficient of relation-
> ship of 1/256, each has relatives in other lineages—
> sister's children, mother's brothers, mother's
> sisters—whose r coefficient is as high as 1/4.

Sahlins goes on to establish that, if we also
take into account the distinction of residence, the
disjunction between genetic and social forms of relationship increases. It is residential groups—which

include both kin (however defined) and non-kin—that are the effective units of social solidarity, cooperation, and resource sharing. Moreover, forms of residence—patrilocal, matrilocal, uxorilocal, virilocal, neolocal—are not always "harmonic" with forms of descent—patrilineal, matrilineal, double unilineal, cognatic—and they may disperse kin from the same descent group across separate residential groupings.

The point is that social arrangements of kinship relations inevitably cut across genetic lines of relationship and sort genetically related people into distinct social and residential groups. Because cooperation and resource sharing are organized in accordance with such social and residential groupings, the logic of social distinctions utterly contravenes that of genetic self-maximization and kin selection—which presumes that one wishes to expend resources on behalf of genetically close rather than distant kin or strangers. These social and residential relations—not the genetic relations emphasized by evolutionary psychologists—are what Sahlins calls the "true models of and for social action. These cultural determinations of 'near' and 'distant' kin... represent the effective structures of sociability in the societies concerned, and accordingly bear directly on reproductive success."

If the argument that one can move from genetic relations to the social relations of kinship does not work for descent-based systems of kinship, it works even less for those systems of kinship groupings that are constituted by reference to other criteria—such as exchange, labor, or feeding. Consider the evidence from the Tanimbar Islands, where I conducted research on kinship and marriage for two and a half years. There, children are allocated to "houses" not by virtue of birth, but rather according to a complex system of exchange that accompanies marriage. If the exchanges are completed, the children are allocated to the house of their father; if they remain incomplete, they are allocated to the house of their mother's brother. Thus, one house may include a diverse array of people who have—as the result of the intricate requirements of exchange—affiliated patrilaterally (to the house of their father) and matrilaterally (to the house of their mother's brother), been adopted as children, or "lifted" into the house as adults. One's "own" genetic children may end up in the houses of other kin or non-kin, while one's own house may be populated by people who either are unrelated or are more distant genetic relatives.

Janet Carsten, one of anthropology's foremost experts on kinship, has shown how, in Langkawi, Malaysia, the mechanism for creating kinship relates ideas of blood to those of feeding. She writes,

> I was told repeatedly that people are both born with blood and acquire it through life in the form of food, which is transformed into blood in the body.... Those who eat the same food together in one house also come to have blood in common.

In this way, fostered children (25% of the children in Langkawi), those related through marriage, and strangers all become kin with those who feed them. Feeding is the means for making kin out of strangers.

Similarly, for the Iñupiat of northern Alaska, kinship is more about "doing" than any essential biological "being." As in Langkawi, Iñupiat kinship is not an essential property given by birth but rather a process through which relations are established and maintained through a variety of means—including sharing food and tools as well as participating together in political and ceremonial events. "Sharing may be both uncalculated and balanced," Barbara Bodenhorn notes, "but among kin it is not dormant. It is *this* labour—the work of being related—rather than the labor of giving birth or the 'fact' of shared substance that marks out the kinship sphere from the potentially infinite universe of relatives who may or may not belong." She observes, "In curious ways, then, 'labour' does for Iñupiaq kinship what 'biology' does for many other systems."

In his classic study, *American Kinship: A Cultural Account*, anthropologist David Schneider demonstrated that kinship relations in the United States are created out of a tension between the cultural value placed on biological relation (defined by such physiological substances as blood or genes) and that placed on the behavioral codes for conduct (defined by such attributes as love and nurturance). In general, the biological is privileged over the behavioral as what makes kinship not only "real" but also distinct from other kinds of relationships. However, there are a number of contexts in which the behavioral codes for conduct come to be seen as that which makes the relationship "real" kinship. Judith Modell has researched the complexities of adoptive relationships among both Euro-Americans and native Hawaiians in the United States. She explains that often, when adopted children in the United States seek their "real" biological parents, they discover that the behavioral part of kinship relations—the years of living together, building up a reservoir of common experience and emotional attachment—is missing, and therefore the sense of relatedness is missing as well. Modell concludes that "when people actually interacted, the thinness of a purely biological relationship became apparent," as did the need to work at relationships—whether grounded in biology or not—to make them "real" kinship relations.

Kath Weston notes that the same has been true for gays and lesbians in the United States, for whom "[d]isclosure [of being gay] became a process destined to uncover the 'truth' of [biological] kinship relations," and rejection by biological parents made it evident that "choice always enters into the decision to count (or discount) someone as a relative." Gays and lesbians have created kinship relations out of choice rather than biology—by choosing to form families out of friends and lovers. In the process, Weston observes, the operative logic shifted from one that assumed that shared biological substance—blood or genes—is what makes kinship last and endure, to one that asserts that kinship is what lasts and endures. It is the "doing" of kinship, not the existence of genetic ties, that makes it real.

There is, thus, a range of criteria that humans use to constitute kinship groupings. While some privilege essential qualities of "being" as that which makes kinship "real," others privilege the qualities of "doing" and "creating." In either case, what comprises the relevant qualities of being or doing varies from society to society. The essential qualities of "being" may be blood and bone, flesh and spirit, semen and breast milk, or genetics as coded information. The qualities of "doing" or "creating" kinship may be forged out of exchange, feeding, labor on the land, ritual sponsorship or religious sacrifice.

Of course, one could always say, as Pinker does, that everyone knows who their "real" kin are and therefore makes the appropriate discriminations in terms of differential nurturance, altruism, and allocation of resources. Even allowing for such a distinction, however, it is clear that the patterns of nurturance, altruism, and allocation of resources follow from specific cultural classifications of kin relations and from particular cultural understandings of appropriate kin behavior. These classifications and understandings are never simply reflections of genetic relation and self-maximization and, moreover, they have significant implications for reproductive success that do not support the arguments of evolutionary psychologists.

What evolutionary psychologists have done is reduce a symbolic, culturally mediated system to what they deem to be a natural, culturally unmediated one. Yet the diversity of cultural understandings of kinship and the range of kinship formations cannot be accounted for as a natural system operating by means of a fixed genetic calculus. The extensive anthropological record tells us that what counts as kinship is cross-culturally variable, cannot be presupposed, and certainly cannot be read directly from some underlying biological reality. Pinker describes the assertion of cultural anthropologists that kinship might be created out of something other than genetic relations as a "myth," which he derides as "an official doctrine" of those he calls the "blank slaters." Yet his assertion does not constitute proof. The empirical evidence plainly shows that people around the world form kinship relations not only by reference to culturally specific understandings of physical substance (which may or may not coincide with our understanding of genetics) but also out of a host of other criteria. Categories of kinship everywhere follow specific cultural logics that always exceed and escape the bounds of any supposedly universal calculation of genetic relations. Ideas about procreation and death, substance and conduct, rank and alliance, nurturance and affection, marriage and exchange, law and ritual are all relevant to the definition of what a kinsperson is, and they shape who gets to count as kin to whom. A theory of kinship based on the assumption that there is a straightforward relationship between genetic "reality" and social categories of kinship is incapable of explaining a multiplicity of other kinship realities such as the ones just noted. And it is these culturally mediated realities—not some imaginary "objective" relationship between people—that shape the structures of exclusion and inclusion, of interest and reciprocity, and of the modalities of behavior that constitute the varieties of kinship around the world.

The Poverty of Individual Self-Interest

Because evolutionary psychologists constitute the social as a precipitate of genetic individualism and self-interest, it necessarily follows that kinship relations—and social relations more generally—must be seen as restrictive rather than expansive. That is, in order to foster the proliferation of one's own genetic endowment, one must limit the "investment" of resources to those closest to oneself and not expend them on those genetically unrelated or only distantly related.

However, what we know about kinship relationships is that they are more often created and maintained precisely to *establish and multiply* networks of social, economic, and political relationships, not to restrict them. That is, people happily expend resources on distant kin and unrelated persons precisely to *bring them into* a network of social relations *that are constituted as kinship*. Social prestige follows not from the restriction of kinship relations and accumulation of resources for oneself and one's closest kin but rather from the expansion of kinship networks and the dispersal of resources across a wide range of social relations. The economic logic of such expansive notions of kinship challenges the neo-liberal economic assumptions that are at the core of evolutionary psychology, since the systematic expenditure of resources on strangers and distant kin—in the effort to make them into close kin—defies the logic of genetic self-maximization that is the hallmark of neo-liberal genetics.

A particularly instructive example of such an expansive understanding of kinship is to be found in the institution of ritual co-parenthood (in Spanish, *compadre-comadre*), which flourished in Europe from the ninth century until the time of the Protestant Reformation and continues to exist (under the rubric of *compadrazo*) in Latin America. In Catholic practice, the ritual co-parent acts as a sponsor to a child when he or she is initiated into the church. The relationship is built upon an analogy between spiritual and biological parentage and a trope of spiritual rebirth, which forms "the basis for the formation of ritual kin relationships through the mechanism of sponsorship at baptism." Sidney Mintz and Eric Wolf note that, from the ninth century on, the number of rituals that required co-parents dramatically increased beyond the original baptismal ritual, as did the number of sponsors and participants (up to 30 in the baptismal ritual alone) who thereby became co-parents of the children and kin to parents who were the subjects of the ritual. This institution—although not based on what evolutionary psychologists would call "real" kinship— nevertheless had very real effects in terms of defining the universe of kinship relationships. Those who were related by bonds of ritual kinship were included within the scope of the incest taboo and therefore were required to marry exogamously—that is, outside the bounds of both biological and ritual kin relations. The effect of this taboo was considerable: the "incest group, biological as well as ritual, was extended to cover seven degrees of relationship." Moreover, the relationship involved material as well as spiritual resources, and "parents attempted to win for their baptismal candidate material advantages through their choice of godparents." Not only did this institution greatly expand the networks of kinship and material exchange beyond those of biological kin, it tended to subordinate "the community of blood to the community of faith" and to privilege the obligations of spiritual kinship over those of biological kinship. As Edward B. Tylor noted of the institution in nineteenth-century Mexico, "A man who will cheat his own father or his own son will keep faith with his *compadre.*"

It is critical to understand that, while the institution of ritual kinship had significant social and material benefits during the feudal era—linking people both horizontally as well as vertically within the social hierarchy—it came under criticism and was ultimately diminished with the Protestant reformation and the beginnings of industrial capitalism. As Mintz and Wolf argue, the

> new ethic put a premium on the individual as an effective accumulator of capital and virtue, and was certain to discountenance the drain on individual resources and the restrictions on individual freedom implicit in the wide extension of ritual kin ties. As a result the *compadre* mechanism has disappeared almost completely from areas which witnessed the development of industrial capitalism, the rise of a strong middle class, and the disappearance of feudal and neo-feudal tenures.

The contemporary forms of the compadre relationship are under similar attack in Latin America today as a result of the increased influence of Protestantism and the ideologies of individualism and capital accumulation.

This points us to the connection between cultural ideologies of kinship and economics. Restrictive kinship regimes—organized by the values of biological relatedness, individualism, and capitalist accumulation—have emerged in particular historical and cultural circumstances and are not the function of a natural and universal genetic logic. In non-industrial, non-capitalist societies, by contrast, kinship relations are, more often than not, expansive rather than restrictive. Strangers are made into kin (and accorded appropriate forms of attention, nurturance, and solicitude) as a way of creating and expanding social relations.

Take, for example, the Nuer, in the era previous to the recent civil war in southern Sudan. The establishment of a man's prestige as a "bull" depended upon a double movement. On the one hand, (genetically close) elder and younger brothers and half-brothers often went their separate ways in an attempt to establish positions independent of one another. On the other hand, any man who wished to become a bull and establish a position of power did so by gathering around himself a host of dependents who were comprised of men from other lineages, tribes, or ethnic groups who were adopted into the group or who married into the group. In this way, close kin were often separated while the relations of kinship and prestige were forged with strangers who were ultimately made into kin through common residence, sharing of resources, and/or marriage.

Likewise, men who occupy the noble houses in the Tanimbar Islands gather around them unrelated men in commoner houses that become affiliated to them as "elder-younger brothers they treat each other well" (*ya'an iwarin simaklivur*). The members of these noble and commoner houses are genetic strangers to one another and distinguished from "true elder-younger brothers." But this does not mean that they are *less* obliged to behave in ways appropriate to kin; on the contrary, it means that they are *more* obliged to do so. Since the relationship exists in, and only in, a conventional understanding that they are mutually obliged to "treat each other well," it is a relationship of kinship that, if it is to exist at all, must be performed through explicit and ongoing acts of nurturance and solicitude. Indeed it is precisely these acts of nurturance, care, solicitude and allocation of resources that turn the relationship into one of kinship.

From the perspective of evolutionary psychology (and of Western regimes of genetic individualism and capitalist accumulation), the expenditure of resources on those who are genetically unrelated or distantly related is, at worst, seen as a "waste" of both genetic and economic inheritance and, at best, seen as a self-interested ploy of reciprocal altruism. From the perspective of societies for whom genetic individualism and capitalist accumulation are not operative, the expenditure of resources on those who are genetically unrelated or distantly related is the very essence of society and social hierarchy. Indeed, it is seen as a requirement for the generation of life and well-being.

It is not that such societies do not recognize radically self-maximizing behavior. On the contrary, they often call it witchcraft. Witches are people who act to restrict their social relations and to accumulate resources for their own individual self-interest. The values that inform evolutionary psychology's account of social life would be seen, from the perspective of these societies, as a manifesto for a society of witches. This is to say that many other societies value social relationship over the individual, exchange over self-maximization, and expansive relations over restrictive ones.

This evidence should alert us to the fact that different cultures constitute and value the individual, society, and the socio-economic modalities that relate them quite differently, and that their relationship can not be presupposed. Evolutionary psychologists would dismiss accounts of cultures in which social and ritual relations are given precedence over individual self-maximization—whether economic or genetic—as the delusion of what Pinker calls the "intellectuals" (thereby apparently claiming the ground of the "anti-intellectual" for himself).

This fundamentalist dismissal of other human realities allows evolutionary psychologists to take a culturally specific understanding of the nature of kinship and social relations—one that emerged under the historical conditions of early capitalism and continues to be informed by neo-liberal economic values today—and transform it into a cross-cultural universal. The evidence, however, makes it clear that it is not a cross-cultural universal but rather a culturally specific understanding whose universality evolutionary psychologists presuppose in their ignorance of the varieties of human sociality.

The Futures in Cloning

From a perspective shaped by neo-liberal genetics, cloning is a dream come true. Since cloning is a form of reproduction that does not require genetic contributions from anyone except one's self, cloning makes it possible to maximize one's own genetic endowment to one's heart's content. I end this section on the individual and society with an examination of a text, "The Demand for Cloning," that imagines the future economic market in cloning from the perspective of evolutionary psychology's neo-liberal genetics. I do so for several reasons. First, this text takes the next—and perhaps final—step in imagining the dissolution of society into individual genetic competition guided by the invisible hand of capitalist market forces. Second, it is an excellent example of the dialectic interchange of economic and biological metaphors that has characterized the fields of both evolutionary and economic theory for over a century, such that each reflects, supports, and justifies the assumptions of the other. Third, it provides an indication of how influential sociobiology and evolutionary psychology have become in the development of a wide range of other academic and applied fields—here the field of legal scholarship. And finally, it is a fascinating example of how a vision of past history shapes a vision of future possibilities.

As soon as cloning became a real practical and technological possibility, its significance for a system of neo-liberal genetics was immediately worked out in an article entitled "The Demand for Cloning" in the edited volume, *Clones and Clones*, and written by Posner and Posner—that is, Richard A. Posner and his son Eric A. Posner, who, like his father, is a law professor at the University of Chicago. Richard Posner, the father, is described in *The New Yorker* as "the most mercilessly seditious legal theorist of his generation... a judge on the Seventh Circuit Court of Appeals... [and] one of the most powerful jurists in the country, second only to those on the Supreme Court."

The elder Posner is a prominent proponent of the application of neo-liberal economic theory, as well as sociobiology and evolutionary psychology, to the resolution of issues relating to the law and jurisprudence. It is, therefore, perhaps not surprising that he would have been drawn to cloning as an issue through which to explore the interrelation between theories of biology and economy.

Building upon the basic assumption of evolutionary psychology that the purpose of human life lies in genetic self-maximization, the Posners postulate that one would expect to find an "evolved preference"—that is, a preference that originated in what is assumed to be the original Pleistocene environment of evolutionary adaptation—for two high-yield genetic multipliers, sperm donation and cloning (but not, evidently, egg donation!). With stunning legal acuity, however, they come to the following conclusion as to why there is not a general rush to donate sperm or clone oneself: "Since there were no sperm banks in the period in which human beings evolved to their present state, a proclivity to donate to such banks has never evolved. Likewise there is no innate proclivity to clone oneself...." They console themselves by observing that—unlike our aversion to snakes, which they presume must have developed in the presence of prehistoric reptilians—no *aversion* to either sperm donation or cloning could have developed, given their absence in the Pleistocene. This, they reckon, is fortunate for the futures in cloning.

The theory that human life exists to maximize the perpetuation of selfish genes is rescued by a second presupposition: that the lines of kinship follow a calculus of genetic closeness. If this calculus is not served by an innate evolved preference for cloning, as such, cloning is served by a supposedly innate evolved tendency to narcissism.

> This narcissistic tendency, which we call evolved rather than acculturated because of its universality and its importance to reproductive fitness—people who don't have a strong preference for their own children are unlikely to produce many descendants—is likely to make some people, perhaps a great many people, desire perfect genetic copies of themselves.... This preference would be a logical extension of the well-documented tendency in animal species and primitive human communities to assist relatives in proportion to the fraction of shared genes. That proportion reaches 100 percent for clones and identical twins.

As far as possible from an exchange theory of social life, the Posners generate social life out of pure narcissism and genetic self-interest.

In order to answer the central question—"why share your genes if you don't have to?"—the Posners make a third presupposition: that there are "good" and "bad" genes, which, moreover, are presumed to be transparently self-evident, readable, and uncomplicated by environmental factors. In their first model, "[g]ood genes... are positively correlated with worldly success," by which they mean wealth. Indeed, they give themselves a genetic hierarchy that stretches from "good genes" at the top to "bad genes" at the bottom. Although, in their first model, good and bad genes are simply equated with more and less wealth, in subsequent refinements of their model, genetic and financial endowments are separated and made to articulate with one another. The problem then becomes how to maximize one's position on this hierarchy—by hoarding one's good genes (and wealth), by trading wealth for improved genetic endowment, or by trading good genetic endowment for wealth.

Once they posit a disjunction between wealth and genetics, it is impossible to *improve* one's endowment on either score without marriage and/or sexual reproduction. Thus marriage and sexual reproduction, they argue, "will appeal especially to people whose success in life exceeds what one would have predicted from knowing their genetic endowment [although how one would know anyone's genetic endowment is unclear]. These people can 'buy' the superior genes of a spouse with the financial resources or social prestige that is the fruit of their worldly success." But it will also work the other way around. "People with good genes but little wealth would want to 'trade' their genes for money in order to have the wherewithall to support and financially endow their offspring, while wealthy people with poor genes would want to trade their money for genes." So at those points in the hierarchy where financial and genetic wealth diverge, marriage and/or sexual reproduction would be in order, whereas at those points in this imagined hierarchy where financial and genetic wealth are matched, there would be a demand for cloning. The Posners assert that cloning "would benefit mainly wealthy women with good genes and to a lesser extent wealthy men with good genes. One would therefore expect... a growing concentration of wealth and highly desired heritable characteristics at the top end of the distribution of these goods and fewer marriages there" [at the top].

There are other reasons, besides trading relative genetic and financial advantage, however, that people might want to marry, according to the Posners. People might want to share rather than conserve their genetic endowment because this is the "price of marriage." That is, if you "put a high value on marriage or the particular marriage partner," "you will have to give your spouse a share of 'your' children's genes." Or, one might want to share or "sell" one's genetic endowment in order to secure an "altruistic" motivation from one's spouse to care for one's children. The problem is this: if the partners in a married couple each cloned themselves, their spouses would not be related to their children. If one assumes, as the Posners do, that kinship behavior is elicited only in proportion to genetic relatedness, then "each spouse may have difficulty thinking of himself or herself as a parent of both children; so dual cloning may not produce dual parenting." This would change under the regime of sexual reproduction (note the shift in the gender of the agent):

> The man who "sells" his wife a genetic half-interest in "his" children gets in return more than someone who will take a share (maybe the lion's share) in the rearing of the children. He gets a child rearer who has a superior *motivation* to do a good job precisely because of the genetic bond. Altruism is a substitute for market incentives, and the man can take advantage of this substitute by giving his wife a genetic stake in his children.

Never mind that altruism, in this passage, appears to substitute for market incentives only in women, or that a woman who "sells" her husband a genetic half-interest in "her" children is, at least in the United States, unlikely to get an equitable share (let alone a lion's share) of childrearing in return. In any case, it is hard to see this particular form of altruism as a substitute for market incentives, since it has been produced through the buying and selling of one's children to one's spouse.

For the Posners, social hierarchy (measured in wealth) should follow from genetic hierarchy; but when it does not, it is generated through a calculus of self-maximizing trades that aims to advance one's position in the hierarchy of genetic and financial worth. Marriage constitutes an extremely weak social link that is barely capable of binding two people together, and it does so only on the condition that they "sell" one another a share of their genetic futures in their "own" children. The Posners' account of the market dynamics they imagine will determine the futures in cloning allows the organizing neo-liberal values of evolutionary psychology—self-interested narcissism, genetic and economic maximization, and the dissolution of social life into the workings of the market—to shine through with splendid clarity.

The Posners' account of cloning provides the most recent example of the metaphoric exchange between the fields of biology and economy. Economic self-interest has been, from the very beginning of evolutionary theory, a central metaphor used to conceptualize the biological relations of evolution and the competition at the heart of natural selection. Sahlins quotes a letter Karl Marx wrote to Friedrich Engels, in which he observes:

> It is remarkable how Darwin recognizes among beasts and plants his English society with its division of labor [read, diversification], competition, opening up of new markets [niches], "inventions" [varia-tions], and the Malthusian "struggle for existence."

Once these economic propensities had been naturalized in the biological processes of natural selection, they could be reimported as natural universals to validate the structures of economic relations. Sahlins summa-rizes the dialectic interchange in this way:

Since the seventeenth century we seem to have been caught up in this vicious cycle, alternately applying the model of capitalist society to the animal kingdom, then reapplying this bourgeouisfied animal kingdom to the interpretation of human society.

When the Posners read kinship and social relations through the futuristic tropes of cloning, the two already long-intertwined lines of bio-economic analogies implode upon one another. Here, the rational maximization of economic self-interest is inseparable from that of genetic self-interest and the two are subordinated to the invisible hand of the market and of natural selection, which has been configured as its biological reflection. In the end, society has been dissolved into the workings of the market; and social relations—to the extent that they exist at all—emerge as the result of a genetic calculus that is interested only in the maximization of the self and its nearly indistinguishable economic and genetic endowments.

This kind of neo-liberal bio-economic narrative of our evolutionary past and contemporary nature—the product of a long history of metaphorical exchanges—provides the cultural lens through which our possible futures are envisioned. The limitations of this vision are not an artifact of nature—that is, they are not determined by innate preferences or the logics of genetics and natural selection. Rather, they are an artifact of culture, produced in the context of the increasing cultural dominance of neo-liberal economic values and their reflection in the genetic individualism of evolutionary psychology.

IV. SEX AND GENDER

If the basic premise of genetic self-maximization does not hold up to empirical scrutiny in face of the diversity of human kinship formations, what about the supposedly universal psychological mechanisms that evolutionary psychologists argue characterize male and female evolved psyches?

Evolutionary psychologists begin from two linked assumptions about the gender asymmetry in reproductive investments. On the one hand, because male reproductive investment can be of relatively short duration, the reproductive success of men is presumed to be constrained by their ability to access as many fertile women as possible and to ensure their paternity of the children in whom they actually invest long term. On the other hand, because female reproductive investment is of relatively long duration, their reproductive success is constrained by their ability to access men with resources to support a small number of offspring. This basic asymmetry in reproductive investments, evolutionary psychologists argue, presented distinct adaptive problems for men and women in the Pleistocene. These problems were solved by the development of gender- and content-specific psychological mechanisms that are both universal and innate.

In this section, I do not wish to get bogged down in the plethora of specific psychological preferences proposed by evolutionary psychologists. Rather, I want to consider three presumed universals that form the logical prerequisites for the existence of these preferences: that it is men who control the resources and that both the sexual "double standard" and male proprietariness over females are natural and innate. Does the empirical evidence support the universality of these claims? If not, the entire architecture of psychological modules begins to collapse, and we will need to question the providence of these supposed universals.

Tracking the Resources

As we saw earlier, through the process of reverse engineering, evolutionary psychologists have told an origin story that is meant to establish the universal preference of females for mates with resources. What is problematic about this scenario is not the proposition that women might prefer to marry men who would make a positive contribution to the social and economic resources of a kinship unit, although it hardly seems necessary to resort to innate mechanisms to account for such preferences. Rather, what is problematic is the assumption that it is men who always and everywhere control critical resources. Wright, for instance, simply asserts, without providing evidence, that "males during human evolution controlled most of the material resources." A corollary problem is the presupposition that it is women, but not men, who are concerned with the social and economic resources that their spouses might contribute to the kinship unit. Such a perspective erases the fact that, in all societies, men and women are embedded in a gendered division of labor in which the totality of productive and reproductive tasks are divided between them in specific and complementary ways.

As historians have documented, it is true that, since the late eighteenth century, a separation has emerged between the productive domain of work and the reproductive domain of the domestic unit—that is, between the spheres of males and females—among the middle and upper classes in industrial societies in the West. But this marked a transformation in family structure. During the colonial era in the United States, for instance, the domains of family and community, home and work, reproduction and production were not separated either functionally or ideologically. "Their structure, their guiding values, their inner purposes," historian John Demos notes, "were essentially the same." This changed radically with the industrial revolution, when a clear separation was established between the home and work, reproduction and production. It should be stressed that what Americans now understand as "traditional family values" is a result of this historical transformation and, further, that this historically particular development does not reflect either the contemporary or historical variety in the division of labor cross-culturally.

In most societies, the totality of productive tasks has been, and continues to be, divided between the genders in such a way that each is dependent upon the other for the procurement of the full range of resources necessary to sustain them. As a consequence, men and women are equally disadvantaged unless they are members of a productive unit that includes both parts of the gendered division of labor. The assumption that parental investment is such that women, but not men, are dependent upon the productive labor and resources of their spouses seriously misconstrues the gendered structure of the division of labor in human societies, and particularly in the hunting and gathering societies that evolutionary psychologists argue provide a window onto the environment of evolutionary adaptation.

The scenario of evolutionary adaptation proposed by the evolutionary psychologists relies on narratives that stress the primacy of the hunt not only to the provisioning of the family but also to the rise of language and culture. This myth of "man the hunter" was debunked long ago, and evolutionary psychologists are aware that prehistoric women did not simply stay home and take care of children but also engaged in productive labor of gathering plant foods and hunting small game. However, this awareness has not unsettled the set of assumptions about gender relations that guides their research on "mate preferences." While evolutionary psychologists continually stress women's desire to find mates who are industrious, productive, and laden with resources, they never consider the possibility that men might be concerned to find women with similar qualities.

Yet from the very beginning of hominid history, the most reliable and proportionally larger source of food was obtained through gathering rather than hunting. Indeed, Nancy Tanner and Adrienne Zihlman argue that the innovations in the creation of tools and containers that marked transitional hominid adaptations on the African savanna five million years ago derived not from *hunting* large animals—hunting at this early stage was non-existent—but rather from "*gathering* plants, eggs, honey, termites, ants, and probably small burrowing animals." Instead of presuming that transitional hominid females were burdened by the long dependency of infants and children, Tanner and Zihlman make a convincing case that it was precisely this dependency that motivated not only the invention of tools, containers, and carrying slings that would facilitate gathering with dependents in tow but also early forms of female-centered social organization and sharing. The toolkit necessary for large-scale hunting was not available in the Middle Pleistocene, and "[h]afted tools and wooden spears, which are part of our conventional image of such a hunting technology, do not appear in the archaeological record before about 100,000 years ago." It thus becomes evident that, during the Pleistocene—that mythical era of evolutionary adaptation—gathering was the primary means of subsistence and, where hunting occurred, it was the reliability of gathering that made possible such an otherwise inefficient expenditure of time and energy. For, "in spite of time-consuming behaviors which frequently yield no food, such as hunting or obtaining raw materials from some distance away, individuals engaged in such activities, probably primarily males, could follow these pursuits because they were assured of a share of the food gathered by women with whom they had close social ties." If anyone was dependent upon others for resources in the "environment of evolutionary adaptation," it was more likely men than women. One could therefore invert Pinker's assertion that "men, because they get meat from hunting and other resources, have something to invest" to say that men could engage in unreliably productive hunting expeditions precisely because women had resources to invest in their husbands and children.

The importance of gathering among contemporary hunter-gatherer societies has long been established. According to Zihlman, "[s]tudies of living people who gather and hunt reveal that throughout the world, except for specialized hunters in the artic regions, more calories are obtained from plant foods gathered by women for family sharing than from meat obtained by hunting." More specifically, Patricia

Draper notes that !Kung San women of the Kalahari "are the primary providers of vegetable food, and they contribute something on the order of 60 to 80 percent of the daily food intake by weight." While meat might be a prestige good, it is neither a predictable nor reliable source of food. Thus, !Kung San women supply both a disproportionate amount of food, relative to men, and food resources that are highly reliable and predictable. Since !Kung San women retain control over the resources they gather, it is men who are dependent upon the resources of women for daily subsistence rather than the reverse.

The interlocking, interdependent nature of male and female labor is also evident among the Yup'ik Eskimo hunter-gatherers. As Ann Fienup-Riordan observes, "Men and women work together in the capture and preparation of each animal, yet never duplicate effort. Specific work configurations complement specific subsistence activities" and none of these activities could be completed without the interdigitated labor of both men and women. Take seals, for example. Men hunt and capture seals, and while they may do the initial processing of larger seals, women do the initial processing of smaller seals. More significantly, women cut and dry thousands of pounds of meat (at a rate of 100 pounds a day), render oil from fat, and tan skins for shoes and clothing. Thus, while men may hunt and capture seals, their work would be for naught if the women did not process the meat and oil into a form that can be stored and eaten during the winter and did not fashion food containers, tools, and clothing out of various parts of the seal. Under such conditions, it is hard to imagine that Yup'ik men would not seriously consider the industriousness of potential spouses and their ability to supply them with the resources that are critical for subsistence throughout the year.

The importance of labor in the choice of the spouse—and its implications for marriage preferences—is by no means specific to hunter-gatherer societies. On the atoll of Truk in the Caroline Islands in Micronesia, subsistence activities include fishing, gardening, and fruit tree cultivation. While both men and women look for physically beautiful and sexually compatible spouses, Ward Goodenough remarks, "they look even more for good workers." He goes on to say that a "person incapable of work is not likely to get married. Physical beauty in one's spouse, while desirable, is subordinate to industry and skill."

In the Tanimbar Islands, the gendered division of labor forms the heart of subsistence activities. While men hunt wild pigs, women care for domesticated pigs; while men do deep-sea fishing, both men and women fish and collect shellfish in the reefs; while men cut down and burn trees to establish swidden gardens, women plant, weed, and harvest the gardens; while men carry building materials home from the forests, women carry food supplies home from the gardens and water from the wells; while men thresh the rice, women pound and winnow the rice; while men build houses and boats, women plait baskets and weave textiles. In such a system, specific resources and the types of labor appropriate to them are gendered; and it is the complementary articulation of gender-specific labor and resources that makes subsistence possible for everyone.

Indeed, in Tanimbar, as in many other societies, it is not just subsistence that requires both male and female resources. Marriage exchanges—which both establish the relative permanence of the relation between a husband and wife and allocate children to the group of either the mother or the father—also involve goods that are both gendered and the product of gender-specific labor. "In everyday life and on festive occasions, 'female' wife-takers give the products of male activity—meat, fish, and palmwine—to their 'male' wife-givers, while the latter reciprocate with the products of female activity—garden produce and betel nut." This is not a world where men have resources and women seek to secure them through their relations with men. Rather, it is a world that is envisioned as the productive interchange of male and female resources, which are, in turn, the product of the labor of men and women.

Even a cursory consideration of the division of labor makes it evident that the production and control over resources does not belong solely to men. The idea that men around the world are concerned with the fertility of women but not their control over resources—and that men therefore have innate psychological preferences that compel them to seek women who are young, beautiful, and shapely but not women who are productive, industrious, or dependable—relies upon a serious misrepresentation of the way subsistence resources are secured, processed, and allocated throughout the world. Surely women are concerned to find industrious men who are willing to contribute resources to the family unit; but men are just as concerned to find women with the same qualities. For a hunter who is not attached to a gatherer is a hungry man; a man who catches seals will have a pile of rotten flesh on his hands, and no clothing on his back, if he cannot find a woman who will process his catch; a man who cuts down the forest for a swidden garden has nothing but a patch of burnt ground if there is no woman to plant and harvest the crops; and a man who brings gifts of male labor to a marriage exchange will remain a bachelor unless they are reciprocated by gifts of female labor. Indeed, if men truly chose spouses on the basis of the qualities that evolutionary psychologists purport to be universal male preferences, they would be at a significant disadvantage in terms of their own survival, not to mention that of their progeny. In the end, evolutionary psychologists have imagined a reproductive strategy for men that would be highly unadaptive in most societies and across most of human history. Although it reflects the division of labor in industrial societies, it does not accord with or account for the diverse patterns of the division of labor found in the world.

The Oxymoronic "Male Sexual Mind"

If evolutionary psychologists assume that the primary problem of women is to find men with resources to "invest" in their relatively few offspring, they assume that the primary problem of men is twofold. On the

one hand, men are concerned to find fertile women—
and preferably to mate with as many as possible in
order to maximize their genetic offspring. On the other
hand, given the perpetual uncertainty of paternity, men
are concerned to find the means to ensure the fidelity
of their spouses, since it is presumed that they will not
wish to "invest" in children not genetically related to
them. Both the double standard and male proprietariness over females are assumed to be solutions to the
adaptive problems faced by males, and therefore to be
natural and innate.

The basic proposition is that males—unlike
females—want as much sex as they can get with as
many partners as possible. Pinker asserts that the "male
sexual mind" is "easily aroused," has a "limitless
appetite for casual sex partners," and, indeed, an "insa-
tiable" desire for "a variety of sexual partners for the
sheer sake of... variety." Indeed, the "male psyche" is
thought to have an "evolved appetite" for harems and
polygamy and to be inherently antithetical to
monogamy. While women supposedly "love marriage,
men don't," Wright argues—noting that "giving men
marriage tips is a little like offering Vikings a free book-
let titled 'How Not to Pillage.'" He explains that the
inherent promiscuity of "the male mind is the largest
single obstacle to lifelong monogamy." The discussion
is skewed from the beginning, since evolutionary
psychologists presuppose that the problem to be
explained is the male (not the female) accommodation
to monogamy.

Whether or not men are able to overcome their
promiscuous and polygamous tendencies, their accep-

tance of monogamy is supposed to require the assurance of fidelity on the part of their spouses. What evolutionary psychologists are concerned with is the negative consequences imagined to follow from the promiscuity of a woman: that is, that the man might be cuckolded and deceived into "investing" in children that are not genetically his own. Thus, long-term investment in women and children is understood to come only in exchange for a guaranteed certainty of paternity, which requires strict fidelity on the part of the woman.

In all of this, the evolutionary psychologists unabashedly write what they call "the Madonna-whore switch" into the genetic hardwiring of the "male sexual mind." Men seek "loose" women in order to proliferate their genes as widely as possible; and they seek chaste or "coy" women to marry in order to be able to guarantee the paternity of the children they "invest" in (thereby assuring that their financial and genetic "capital" flows through the same veins). Indeed, this double standard is seen as the "optimal genetic strategy" not only for all human males but, more fundamentally, "for males of any species that invest in offspring: mate with any female that will let you, but make sure your consort does not mate with any other male." The double standard, therefore, is understood not only to be a precipitate of the supposedly natural logic of genetic self-maximization but also to be realized through a set of preference mechanisms that are presumed to be genetically encoded in the male psyche.

This "natural" logic of the double standard generates, in turn, a "natural" logic of the differential disbursement of resources. The assumption is that men will refrain from investing long-term resources in "loose" women (never mind the abundant evidence of men who have lavished untold wealth and attention on their mistresses), reserving those resources for those otherwise chaste women whose children's paternity can be guaranteed. Wright notes that this "leads men to shower worshipful devotion on the sexually reserved women they want to invest in—exactly the sort of devotion these women will demand before allowing sex. And it supposedly lets men guiltlessly exploit the women they don't want to invest in, by consigning them to a category that merits contempt." The Madonna-whore distinction thus appears to sort women into two natural kinds, whose essential qualities men's psychological mechanisms unconsciously perceive and deem worthy of either "investment" or "exploitation."

As an aside, it should be said that evolutionary psychologists are clear that the ruthless efficiency of this "natural logic" follows male not female interests. They seem singularly unconcerned with the consequences of male promiscuity for women and children who are abandoned by philandering men—and for whom this "natural logic" is hardly an efficient adaptation. But, in the minds of evolutionary psychologists, "loose" women—presumably less well adapted, with less discriminating psychological mechanisms—appear to deserve their fate.

Like the double standard, male proprietariness over females is assumed to be a natural, evolved characteristic of the "male sexual mind." As Wilson and Daly put it: "men lay claim to particular women as songbirds lay claim to territories, as lions lay claim to a kill, or as people of both sexes lay claim to valuables." Since female infidelity is seen as a "threat to male fitness," male proprietariness in relation to women is thought to evolve naturally from men's need to avoid cuckoldry and ensure female fidelity. The forms of male proprietariness, evolutionary psychologists argue, include a wide range of activities—veiling, various forms of claustration, bridewealth, and adultery fines—all of which presumably indicate the ways in which women become the property of men or are turned into commodities traded by men.

The double standard and male proprietariness may seem "self-evident" to Americans. But it is methodologically invalid to assume that what *feels* self-evident and natural from one's own cultural perspective must therefore *be* natural and universal in all cultures. It is imperative that we ask just how universal they are and whether they can actually account for the forms of sexuality found the world over. To query the presumed universality of these ideas, I examine three aspects of the argument: that men everywhere seek to multiply their sexual contacts with women; that the double standard is both natural and universal; and that male proprietariness over women is as natural and inevitable as a lion's proprietariness over its kill.

The Cultural Values of Promiscuity

The incidence of promiscuity in men and women varies cross-culturally. This variability is explicable not by reference to a presumed universal logic of gendered sexuality but rather by reference to the particular cultural logics that organize sexuality in different societies. We must therefore examine specific cultural ideas, beliefs, and practices about what constitutes gendered persons and bodies; about the perceived attributes of bodily substances (for instance, semen, blood, or milk); about the processes that account for reproduction, life, and death; about the relation between sexuality and religion, politics, and economics; and about the structures of hierarchy and power, among other things. The relative promiscuity of men or women across different cultures will follow from these understandings.

The dynamics of sexuality among the Etoro, in the highlands of Papua New Guinea, for instance, follow from their conception of *hame*, an essentially formless life force that is manifested in the breath and understood as the animating spirit of all humans. Raymond Kelly, who carried out field work in this New Guinea society, informs us that an increase in *hame* is associated with growth, strength, and vitality; a decrease in *hame* is associated with labored breathing, coughing, chest pains, general weakness, aging, and ultimately death. Because *hame* is especially concentrated in semen, the dynamics of life and death are intimately associated with the loss and gain of semen through both heterosexual and homosexual intercourse. Heterosexual intercourse is necessary since *hame* (and the semen that contains it) is understood to be required for the initial growth of the child in the woman's womb. Homosexual intercourse is necessary since pre-pubescent boys are thought to lack an inherent source of semen and to require an increased supply of *hame*'s vital energy in order to grow and mature. Thus, the insemination of boys by adult men (usually their sister's husband) from the age of about ten until their mid-twenties is seen as a form of nurturance that ensures that boys grow, become strong and healthy, and have the requisite life force to engage in their own acts of reproduction and nurturance.

According to Kelly, the central paradox and tragedy of Etoro sexuality is that, as adult men give life through heterosexual intercourse and foster the growth of boys through homosexual intercourse, they are thought to deplete their own life force and health and bring themselves closer to their own demise. Giving life ultimately engenders one's own death. Not surprisingly, then, there is a kind of moral economy of semen transfers; and, within this moral economy, heterosexual and homosexual intercourse are differently valued. When men consider homosexual (oral) intercourse, the focus is upon the life-giving side of the equation, whereas, when men consider heterosexual (vaginal) intercourse, the focus is upon the depletion of men's *hame* and their consequent enfeeblement and ultimate death. Heterosexual intercourse is surrounded by a wide array of taboos that constrain its incidence: it can not be carried out in the longhouse or its vicinity, in a garden dwelling, or in the garden itself, lest the crops wither and die (which contrasts sharply with homosexual intercourse in the gardens, which is assumed to cause crops to flourish and increase in yield). Heterosexual intercourse—seen as a "fundamentally antisocial behavior"—should only take place in the forest, although, even there, a couple is endangered by the presence of death adders, which are thought to dislike the odor of intercourse. Moreover, heterosexual intercourse cannot be done at various points in the productive cycle of gardens. As a consequence, heterosexual intercourse is tabooed anywhere from 205 to 260 days a year, thereby confining the possibilities for its occurrence to about one third of the calendar year. Both the timing of births and the generally low birth rate appear to confirm the efficacy of these restrictions on heterosexual intercourse and its implications for reproduction.

We are in the presence, then, of a system in which heterosexual intercourse is considered with deep ambivalence and is surrounded with massive restrictions that limit its occurrence and its perceived enervating consequences for men. Moreover, not only are Etoro men ambivalent about engaging in heterosexual intercourse, they also divert much of their vital energies (and, for the evolutionary psychologists, genetic potential) to the insemination of young boys— hardly an optimal strategy for advancing their reproductive success.

Ultimately, the Etoro clearly see promiscuity as a negative value, in that it is associated with the unnecessary depletion of *hame* and the implications of that for the health and very life of the men concerned. Those who unnecessarily deplete men of their semen and *hame* are considered dangerous and worthy of contempt. These include women who make excessive sexual demands on men; young men who have sexual relations with other young men (and therefore deplete their resources before they have been properly established); and witches, who intentionally sap men of their *hame* and its life force. The Etoro moral economy thus challenges the idea that male promiscuity is either simply hard-wired or in the service of genetic proliferation. Indeed, it is hard to imagine a case that throws more obstacles in the way of genetic proliferation.

But then consider what the historian, Ben Barker-Benfield, has called the "spermatic economy" that organized male sexuality, at least in some circles, in the United States in the nineteenth century. Here, too, creative energy was a limited good that circulated within a kind of zero-sum moral economy of which sexual reproduction and social production were both a part. Any form of excessive sexuality—including frequent heterosexual intercourse, masturbation, not to mention homosexual intercourse—sapped men of their energy, and deprived them of the ability to transform sexual energy into the forms of social productivity that allowed them to compete effectively within the larger political and economic domains. The assumption that there was a limited amount of sexual energy and, further, that sexual and productive activities were in an inverse relation to one another, had several implications. If one "spent" one's sperm in excessive sexuality, one would not only have weak, debilitated children but would also be incapable of socially productive activity and end up in a state of nervous exhaustion if not insanity. It was necessary, therefore, to hoard one's sperm and to keep from spending it recklessly. Sperm that had been "saved" was strong, undiluted, and therefore "rich," and was necessary not only for proper reproduction but also for productive activity in society. Thus there was clearly a moral sanction that constrained all forms of sexuality (in some manuals, heterosexual intercourse should properly occur only "at high noon on Sundays"!) in order to harness a man's spermatic wealth to the goal of producing social wealth on the one hand and strong children on the other. Surely not everyone followed all the strictures found in the manuals of proper sexual hygiene. Yet, they comprised a dominant system of ideas that linked sexual and social production in a zero-sum economy of inversely related expenditures of energy. This economy privileged the "saving" over "spending" of spermatic riches as well as the transformation of spermatic "wealth" into social (rather than genetic) wealth.

Taken together, the Etoro and American examples indicate that the narratives of the evolutionary psychologists have been unable to comprehend how the moral economies of spermatic transfers that organize sexuality in various cultures might utterly contravene their presuppositions about the promiscuity of men. Other interests—those of life and death, or competition over the production of social wealth—transform the prospects of promiscuity and the consequent expenditure of sperm into a negative social value and a positive risk to health, life, and wealth. Paradoxically, evolutionary psychologists are unable to foresee how even a self-interested spermatic regime might work against their presuppositions about the primacy of a self-interested genetic regime.

In the Etoro and United States examples, sexuality is shaped by its place within a larger system of moral values that relate to ideas about life, death, and sexual and social productivity. The fact of the matter is that the value of sexuality can never be determined in isolation from other social values. Take, for instance, the power that the religious value of transcendence has to severely compromise the reproductive potential of men.

In Thailand, men ideally spend a portion of their lives in the Buddhist monkhood. Indeed, while scholars estimate that "over half of the eligible Thai males" actually do so, Thai national statistics go even further to claim that "over 95 per cent of eligible males over 50 have served for a time as a monk." Men thus remove themselves from the constraints of material and social life entirely and place themselves in monasteries for varying periods of time before returning to social life and the possibility of heterosexual encounters. The impetus for men to join the monkhood, anthropologist Thomas Kirsch tells us, derives from a dominant set of religious ideas that value the spiritual over the material, and relative detachment over attachment to worldly affairs—including those of the economy and sexuality. Both economic and sexual relations are associated with women and are understood to interfere with the more spiritual goals (ultimately, of attaining nirvana) that men strive to achieve. Thus the shape of male sexuality in Thailand —as in other religiously oriented cultures—can only be determined by understanding its place within a religious system that gives higher value to spiritual and otherworldly concerns and devalues aspects of the material world, including sexuality.

The power of specific cultural ideas and values to organize human sexuality is particularly evident with the Kaulong, a Papua New Guinea society in which Jane Goodale conducted research on gender roles. For the Kaulong, the idea of female pollution is particularly salient. Although women are considered to be continually in a state of pollution, men can protect themselves from the everyday pollution of women by avoiding being directly under a place where women have been or are in contact with objects a woman has touched. However, pollution is intensified during menstruation and childbirth, which both require the lateral separation of women into menstrual and childbirth huts. Sexual intercourse—which is synonymous with marriage—is thought to be particularly polluting for men and to threaten their health and their very life. As a consequence, Goodale reveals, men are deeply fearful of sexual intercourse, avoid it for as long as possible, and typically marry as late in life as possible (saying "I am too young to get married and die"). It is therefore women who aggressively court men rather than the reverse: they give men gifts, engage in playful talk, use magical substances to keep men from fleeing, and even aggressively attack men with knives or switches. Indeed, a pattern of female aggression toward males is established in early childhood, and males are trained not to return any violence rendered them by females.

In the end, men succumb to the advances of women because their fear of death from pollution is outweighed by the cultural imperative to achieve immortality and continuity of identity through the generation of children who will replace them. Any possibility of male promiscuity is virtually obliterated not only by the power of ideas about the lethal pollution of women but also, at least in the pre-contact period, by a series of sanctions that equated sexual intercourse with marriage; prohibited divorce, adultery, and sexual intercourse outside of marriage; and punished offending males (and sometimes females as well) with death or exile.

Such examples, which could be greatly multiplied, suggest that that there is no such thing as *"the* male sexual mind"—promiscuous or otherwise. Rather, the varieties of male and female sexuality are organized by particular cultural ideas and values about bodies, bodily substances, and gender. And, moreover, they are informed by cultural understandings about the place of sexuality in relation to ideas about life and death, and about the value of sexuality relative to other goals of human life. Thus ideas such as those concerning pollution, or the depletion of life force, or the demands of spiritual transcendence all reduce the overall incidence of sexual intercourse in a given culture, either by prohibiting sexual activity altogether or directing it toward ends that do not enhance reproductive success. Evolutionary psychologists may dismiss such ideas as being epiphenomenal to a more fundamental logic of genetic maximization or natural male promiscuity. But the fact of the matter is that the forms and frequencies of sexual intercourse found cross-culturally are always subject to particular cultural logics and can never be fully accounted for by reference to the supposedly universal and natural the reproductive logic of evolutionary psychology.

Unwiring the "Madonna-whore Switch"

In order to establish the universality of the double standard and the existence of an inherent "Madonna-whore switch," evolutionary psychologists would have to demonstrate that men everywhere distinguish between "loose women," whom they feel entitled to shamelessly exploit, and "chaste and coy women," whom they revere, marry, and are willing to "invest" in. The theory presupposes that, universally, women (but not men) who engage in premarital sex are devalued and denigrated as marriage partners while women who abstain from premarital sex are highly valued and sought after. Indeed Wright asserts that there "is a virtual genetic conspiracy to depict sexually loose women as evil." The problem with the theory is that there are numerous societies in which premarital sex is, itself, a highly valued activity for both women and men, and women's engagement in premarital sex does not damage but rather *enhances* their marital prospects.

Since hunter-gatherer societies feature as the prototype for the social environment of evolutionary adaptation in the accounts of evolutionary psychologists, we begin again with them. Signe Howell, a

Norwegian anthropologist who has worked with the hunting and gathering Chewong people of the Malay peninsula since the 1980s, notes simply that a young Chewong "couple would meet secretly for brief sexual encounters in the forest over a period of time before they began publicly to cohabit." These premarital sexual encounters are not stigmatized by these hunter-gatherers. They are seen as the obvious prelude to marriage rather than its antithesis. Indeed, "their term for marriage, or married, is just 'sleep together' (*ahn nai*)."

For the !Kung hunter-gatherers of the Kalahari desert in Botswana, children are accorded significant autonomy, and they go off in groups to set up play villages at the edge of the adult villages. There they would imitate the activities of their parents, including hunting, gathering, cooking, eating, and sexual intercourse. Indeed, Marjorie Shostak's exten-sive interviews with the !Kung woman, Nisa, revealed that !Kung children engage in a considerable amount of sexual play from an early age, play that eventually matures into regular sexual intercourse between unmarried couples. Married life, for most young women, begins with a series of provisional or trial marriages with older men. Due to the age difference, these marriages are rather unstable and easily break up in divorce, which is, as a consequence, quite common. Shostak observes that "[n]o premium is placed on virginity—indeed, I could not find a word for virginity in the !Kung language. The divorced girl or woman simply re-enters the category of highly desirable potential wives, to be sought after by eligible men."

Among these hunter-gatherers, women's extensive premarital and successive marital sexuality has no consequences for their subsequent marital status.

Full adult status for Pokot men and women in west central Kenya depends upon three attributes: sexual skill; circumcision at initiation; and marriage and the birth of children. Robert Edgerton tells us that Pokot attach great importance to physical beauty and adornment, to the development of sexual skills, and the pursuit of amorous adventures. Beginning around ages ten or eleven, Pokot girls and boys begin to engage in extensive sexual play—including dancing, conversation, gift giving, petting, and sexual inter-course—which is seen as critical to developing the skills necessary for a successful marriage, that is, to give one's partner sexual pleasure. Premarital sex for both women and men is seen as a prerequisite for marriage, not an alternative to it. The critical distinc-tion is not between women who have or have not had premarital sex, but rather between women (and men) who have been circumcised and therefore who are able to be married and have children and those who are not. Full adult status comes only with circumcision and, ultimately, marriage and the successful reproduc-tion of children.

A similar evaluation of premarital sex existed in the Melanesian archipelago of the Trobriand Islands. Bronislaw Malinowski, one of the founders of anthro-pology, wrote numerous books on the Trobriand Islands including one devoted entirely to Trobriand sexuality. He tells us that, from a young age, children engaged in all manner of sexual play, including attempts to imitate sexual intercourse, which their parents found amusing. After puberty, sexual liaisons became more serious and sustained (although not entirely exclusive) and special houses were built in which several young couples each had their own private sleeping spaces where they were allowed to develop their sexual relations. Contrary to western conventions, for unmarried couples in the Trobriands, sexual relations were expected while eating together was absolutely forbidden. Whereas young men and women might have a number of different partners before marriage, eventually a relationship with one partner matured into an enduring one that would result in marriage. In addition to opportunities for sexual relations afforded by the existence of special houses dedicated to unmarried couples, there were numerous occasions in which pre- and extra-marital sexual relations might be pursued. Sexual relations often followed from the erotic character of social promenades (*karibom*) and of dances that occurred at harvest festivals (*milamala*) and at ceremonial food distributions (*kayasa*). In addition, groups of boys as well as groups of girls might visit neighboring villages in pursuit of amorous adventures. It is hard to imagine a "Madonna-whore switch" operating in such a culture, since, for both women and men, sexual engagements were sanctioned and facilitated while sexual abstinence was simply not a culturally relevant category. Women were assertive and even aggressive in pursuit of sexual encounters, their sexual exploits were as positively valued as those of men, and such exploits had no negative consequences for their marital status.

Prior to their conversion to Catholicism in the 1950s, Nage and Keo women on the island of Flores in eastern Indonesia participated in a system of premarital sexuality that was a little more formalized than the Trobriand one. Gregory Forth reports that young unmarried women could enter into a temporary sexual relationship—anywhere from one night to several months—with either married or unmarried men. These relationships were common, positively valued, publicly sanctioned, and took place in the house of the girl's parents. The relationship was sanctioned by an exchange of goods that resembled, in form, those made at marriage: the man gave livestock and metal valuables; the woman's parents slaughtered a pig for a final meal between the parents, the woman and the man, and they also presented a textile to the intermediary who had arranged the match. In addition, over the course of their relationship, the man would bring a considerable number of other non-obligatory gifts as a token of his respect. Most young women participated in this form of sexual relationship and took anywhere from a couple to dozens of lovers. While casual sex (that is, sex that occurs without parental approval) was not sanctioned,

> the status of mistress (*ana bu'e*) is described as having been an honourable one. Such women were not stigmatized in regard to their later marriage to other men. Being a mistress was never an alternative to marriage, as it has sometimes been in the modern west.... If anything, formally contracted premarital sexual affairs were viewed as a prelude to marriage, for women and men alike.

While evolutionary psychologists might figure that this institution of publicly sanctioned premarital sex worked to the advantage of men—providing a means of proliferating their genetic endowment, it was not shaped by a dichotomy between "loose women" whom one shamelessly exploits and devalues and "coy women" whom one honors and "invests" in. Indeed, a Nage or Keo man respected and honored his mistress, was required to "invest" in her and her family in order to engage in the relationship, and might subsequently marry her (or another woman who had been the mistress of other men). Moreover, a man had no inherent claim over any child that might issue from the relationship: such a child belonged to the woman's "house" (*sa'o*) unless he "invested" further, by making a prestation of valuables to secure the affiliation of the child to his own "house." From the woman's perspective, such relationships had plenty of advantages: the woman gained in prestige; developed romantic relationships; was, with her family, in receipt of significant prestations; and had the opportunity to enjoy "sex, or what might even be called 'free sex', that is, physical relationships free of the considerable obligations entailed in the statuses of wife, daughter-in-law, and mother."

Cases of polyandry pose a particularly interesting challenge to the universality of the Madonna-whore distinction, since they positively sanction women who take on multiple sexual and marriage partners. In the central kingdoms of Calicut, Walluvanad, and Cochin in the center of the Malabar coast or Kerala in India in the period prior to 1792, commoner

Nayar girls aged 7-12 underwent pre-puberty marriage rites in which they were ritually married to men of their own sub-caste from linked lineages—a ceremony that allowed them, subsequently, to take on multiple visiting husbands. As Kathleen Gough describes it,

> The ritual bridegrooms were selected in advance on the advice of the village astrologer at a meeting of the neighbourhood assembly. On the day fixed they came in procession to the oldest ancestral house of the host lineage. There, after various ceremonies, each tied a gold ornament (*tāli*) round the neck of his ritual bride.... After the *tāli*-tying each couple was secluded in private for three days.

Various rights were conferred by this ritual marriage: the ritual husband had the right to deflower his bride—although this was apparently a task that was "viewed with repugnance" (not the eagerness evolutionary psychologists might have predicted)—as well as the right to be mourned at his death by his ritual wife. The ritual wife had the "right to *have* a ritual husband of her own or a superiour sub-caste before she attained maturity." If a girl had not been married to a man of the appropriate sub-caste before puberty, she would have been ostracized and even murdered.

However, once a woman had passed through the ceremony of ritual marriage, she was free to take on multiple visiting husbands as long as they were from her own sub-caste or higher castes. Reports note that women took anywhere from three to twelve visiting husbands. Although commonly she would have a smaller number of husbands in a more or less long-standing relationship, she "was also free to receive casual visitors of appropriate sub-caste who passed through her neighbourhood in the course of military operations."

It is difficult to locate Nayar polyandry—or any polyandry for that matter—within the evolutionary psychologists' Madonna-whore schema. Relative to monogamous systems, Nayar women were free to engage in short or long term relationships with multiple husbands and lovers. The idea, here, was not that women must be chaste, coy, and save themselves for one man; but rather that they must save themselves for *men of their own sub-caste or higher*. The sanction was not against multiple sexual and marital relationships, but rather against sexual or marital relations with lower caste men—for which a woman might indeed lose her life. Nayar women were not "whores" subject to ruthless exploitation as they engaged in multiple sexual/marital relationships. On the contrary, having been ritually married and having taken husbands of the appropriate sub-castes, they were honorable women and treated as such by their husbands and lovers.

Evolutionary psychologists might counter that the "Madonna-whore switch" is not triggered here because men are not required to "invest" in their own children but only those of their sisters. But such a system tells us that there is not an automatic "switch" that is hard-wired in men: women here are not naturally perceived as "loose" or "chaste" depending upon the *number* of men with which they have relationships. A Nayar woman's honor (or a man's for that matter)

102

depended rather upon other cultural distinctions—specifically those that determined the lineage, sub-caste, and caste of her partner.

A "Madonna-whore" distinction rests upon a particular valuation of virginity and sexuality, of male vs. female sexuality, and of premarital, marital, and extramarital sex. Although some cultures obviously do make evaluative distinctions between sexually promiscuous and chaste women, between male and female sexuality, or between sexuality within and outside of marriage, examples like the ones discussed here demonstrate that many do not. The cross-cultural varieties of male and female sexuality can not be comprehended within the framework of the "double standard" for the simple reason that the ideas and values that inform the "double standard" and the "Madonna-whore" distinction are culturally and historically specific rather than universal.

Like a Lion and its Kill:
Proprietariness and its Discontents

Evolutionary psychologists assume that men do not wish to be cuckolded and thereby be placed in the situation where they would be "investing" in children who are not their own. For this reason, they assume that men are inherently proprietary over their wives—that is, over the women they *do* intend to "invest" in—so as to secure both their wives' fidelity and their own certainty of paternity over their wives' children. I would hardly wish to argue that there is no evidence for male proprietary behavior relative to women. On the contrary, there is plenty. But there are also plenty of examples of cultures in which the idea and practice of male proprietariness is simply non-existent, or where it works, paradoxically, counter to the establishment of certain paternity. What does some of that contrary evidence look like and how might we interpret it?

To begin, again, with the hunting and gathering societies that feature so prominently in the narratives of evolutionary psychologists. Signe Howell argues that, among the egalitarian Chewong of Malaysia, there are no political leaders, no political hierarchies, and no institutionalized forms of authority. The relationship between husband and wife, while structured by conventional expectations about the kinds of roles they will play, is not shaped by differential status or power. In cases of adultery, no fines are paid to the woman's husband by her lover. "That is not our way," people commented. Divorce is fairly common and children may accompany either parent, although they normally follow the mother, when small. These details are significant, since evolutionary psychologists read adultery fines paid to the husband of the woman and male control over legitimate children to be evidence of the natural proprietariness of men.

!Kung men and women value both marriage and the excitement and novelty of extra-marital affairs, which—according to Marjorie Shostak's interview with Nisa—they appear to relish and engage in quite frequently. While such affairs are normally kept secret, when found out, they provoke jealousy, threats of violence, and, occasionally, actual physical violence. However, while men are most often the initial aggressors in fights with their wives, jealousy and the threat or actuality of violence are not at all the sole preserve of men (as the theories of evolutionary psychologists would suggest). Both women and men are capable of expressing jealousy, and of threatening and engaging in violence toward partners or competitors. Indeed, noted ethnographer of the !Kung, Richard Lee, writes that, of the fights he observed between 1963 and 1969, "women [were] involved in fights almost as frequently as men (23 vs. 16 times)," and that adultery "cropped up in 2 (of 11) male-male fights and 2 (of 14) male-female fights, but was a factor in 5 (of 8) female-female fights." However, family and neighbors quickly intervene to stop arguments and fights, most often while they are still at the verbal stage and before they escalate into physical or lethal violence. Neither the data on !Kung violence nor the lack of dowry, brideprice, and adultery fines among the !Kung, provide evidence for male proprietariness over females in this hunter-gatherer society.

In the Melanesian society of Lesu, Hortense Powdermaker reports, people engaged not only in three forms of marriage—monogamy, polygamy and polyandry—but also in both pre-marital and extramarital sexual relations. Generally, a woman would marry shortly after the ritual that marked her first menstruation but, if she did not, it was not unusual for her to engage in a series of sexual relationships before she married. In either case, once married, both men and women were expected to—and, in fact, did—engage in extra-marital sexual intercourse with many partners over the course of their lifetime. Middle-aged women, Powdermaker observes, "had had so many lovers that they could not remember the exact number." Every time a man had sexual relations with a woman who was not his wife, he presented her with a string of shell currency (*tsera*), which she would then present to her husband, who "gladly accepted" the gift, understanding full well where it came from and why. Powdermaker notes that the "*tsera* is not a payment any more than the marriage price is a payment for the bride. It is simply part of the traditional code of reciprocity whereby nothing is ever given for nothing." All the children of a married woman were accepted by her husband as his own, regardless of their paternity. Although, on rare occasions both women and men might express jealousy, this was not at all the norm and was not socially sanctioned. Extra-marital sexual activity was not stigmatized or punished and it was not the cause of jealousy or subject to the double standard. Quite the contrary, it was a valued social practice that men and women engaged in with equal pleasure and frequency.

While the male proprietariness presupposed by evolutionary psychologists involves the entitlements of individual men and the establishment of the paternity of individual men, such individual proprietariness is a structural impossibility in systems of polyandry, where women take multiple husbands. Sometimes these husbands are related as brothers, as in fraternal polyandry, sometimes they may be related only as village mates, or members of the same sub-caste, as in the case of the Nayar.

A Nayar woman's ritual marriage did not establish the individual proprietary rights of the ritual husband over his ritual wife. Rather, Kathleen Gough tells us, they established the rights of a woman to access any and all men in her sub-caste (outside of her lineage). Her visiting husbands had no proprietary rights over her or her children and they could be taken on or dismissed at will. Gough notes that a "husband visited his wife after supper at night and left before breakfast next morning. He placed his weapons at the door of his wife's room and if others came later they were free to sleep on the verandah of the woman's house." Individual proprietary claims over a woman on the part of any man were structurally nonsensical, since the entire system was organized so that the woman had ritual and legal access to a whole category of men.

In order to establish the legitimacy of her children's place in her matrilineage and sub-caste, a Nayar woman was required not only to have been ritually married (before she gave birth to any children) but also to have had the delivery expenses for the birth of her child assumed by one or more of her husbands (if no man stepped forward to pay the delivery expenses, it was assumed that the child's father was a man of a lower caste, and the woman would have been ostracized or killed). From the perspective of evolutionary psychology, we have here a doubly paradoxical situation: not only would multiple husbands acknowledge their paternity relative to a particular child but the purpose of that acknowledgement was to secure the membership of that child in the sub-caste and the matrilineage of the mother, not the father.

The Lele of the former Belgian Congo practiced monogamy, polygyny, and a form of polyandry in which approximately one in ten women became what was known as a "village wife" (*hohombe*). Mary Tew, who is known by her married name as the renowned British anthropologist Mary Douglas, carried out her initial field work among the Lele. She observes that, "[w]hether captured by force, or seduced, or taken as a refugee [from an abusive husband], or betrothed from infancy, the village-wife is treated with much honour." Indeed, she could be the daughter of a chief who was given in exchange to a village, which then assumes a collective in-law relationship with the chief. During the first six months or more of the "honeymoon" period, the village-wife was exempted from the usual division of labor, did not engage in heavy work, and was pampered by the men of the village who even did her "female labor" for her and showered her with gifts of meat and other favors. During this initial period, different men would sleep with her in her hut every two nights, but she could have sexual relations with any village man when she went to the forest. After this period ended, when she was "brought out," she was allotted "a limited number of husbands [up to five]... [who were] entitled to have relations with her in her hut, and to be cooked for regularly." Over time, the number of husbands would decrease, but she continued to be allowed to have sexual relations with any village man in the forest. Paternity of any children of the "village-wife" was claimed by all the men: "All of us. We begot him, all the men of the village.'" These children were honored throughout their life—especially at marriage and death.

The fact that males do not everywhere feel the need to (and are not structurally in a position to) establish exclusive rights over women—in order to maintain their certainty of paternity—is clearly demonstrated in cases of polyandry. Such systems reveal that female coyness, male concern with the certainty of paternity, and *individual* male proprietariness (or even male proprietarinesss in general) are hardly innate psychological mechanisms.

But this is evident not only in cases of polyandry. Other values—whether these are economic or spiritual—may direct men and women toward more inclusive rather than exclusive sexual relationships. Such inclusive relationships may co-exist with male proprietariness, but the latter does not always serve to secure certainty of paternity.

Consider the spouse exchange that was practiced by the Iñupiaq and Yup'ik Eskimos of Alaska, including those who inhabit the area between the Brooks Range and the Arctic Ocean. In this area, Robert Spencer tells us, the value of extending the bonds of productive cooperation beyond those of the extended family resulted in a particular structure of marriage rules, which included exogamy, the rejection of cousin marriage, a prohibition on two brothers marrying two sisters, and prohibitions on a man marrying two sisters in polygynous unions, or a mother and a daughter, or the sister of a deceased wife. Each of these injunctions had the effect of extending and multiplying (rather than concentrating) a group's alliances and, therefore, their relations of cooperation and mutual aid. The solidification of relations of cooperation and mutual aid was also effected through the practice of wife exchange. Although men were normally jealous of their wives' extra-marital affairs, they readily lent their wives to their trading partners, or left them with friends, neighbors, or partners when they went on extended hunting or trading expeditions. In all these cases, as long as the receiving men were not relatives, it was expected that sexual intercourse would occur. The women's sexual relations with both men did not produce a jealous reaction, but, on the contrary, cemented the existing cooperative relationship and extended it into the following generation to the children of both couples. The result, however, was that men willingly placed themselves in a situation in which the paternity of their wives' children was uncertain and they might end up investing in children not genetically their own. In such an environment, evolutionary psychologists might argue, the resulting reciprocal altruism might be well worth the uncertainty of paternity. Yet, if certainty of paternity were the primary issue, there are surely innumerable ways of multiplying the bonds of reciprocity and cooperation without also undermining the certainty of paternity. It is not clear, for instance, why wife lending is necessary to cement the already strong bonds of trading partners, friends, and neighbors. The point is that the cultural value of reciprocal cooperation is given priority and is expressed through forms of extra-marital sexuality that muddle paternity, produce friendship rather than jealousy, and inevitably result in men caring for children that are not genetically their own. Again, the presuppositions of evolutionary psychology are unable to predict that even where male proprietariness may be in force, as among the Eskimo, it will not necessarily operate in the service of individual self-interested genetic proliferation. While proprietary rights over wives exist in this case, they do not establish exclusive sexual relations but rather make possible a wife's intercourse with her husband's partners, friends, or neighbors—a practice that confuses paternity at the same time that it fails to provoke the jealousy evolutionary psychologists assume would naturally follow.

A similar effect resulted from the ritual practices of the Marind-anim of Papua New Guinea. These practices are the subject of a massive volume, *Dema: Description and Analysis of Marind-Anim Culture*, written by the Dutch civil servant and ethnographer J. van Baal, who lived in the area for two years. In this society, a form of sexual license was a standard part of a wide range of rituals. A bride was expected to engage in sexual intercourse with five to ten of her husband's clanmates on her wedding night and also at those points where she emerged from the prohibitions on sexuality that surrounded birth. Several women were similarly expected to engage in acts of sexual intercourse with multiple partners on virtually every ritual occasion—which included everything from age-grade feasts to funerals, from feasts celebrating the preparation of new gardens to those held in connection with big hunting and fishing parties, from feasts promoting the fertility of crops to rites dispelling sickness. This abundance of sexual intercourse served a number of purposes, which primarily derived from Marind-anim understandings of sperm, which was seen as "the essence of life, of permanence, of health and prosperity." First and foremost, Marind-anim understood that the fertility and health of both humans and crops required the production of an abundance of sperm. Hence the ritual insemination of a woman by the community of her husband's clanmates at the time of her marriage and at the end of her confinement had as its purpose the enhancement of the woman's fertility and the fulfillment of her reproductive potential. Similarly, the fertility of crops and the productivity of large hunting and fishing expeditions were enhanced by the production, collection, and magical use of quantities of sperm. Sperm was also understood to have medicinal value, and was collected through multiple acts of ritual intercourse with women and used in healing—being both ingested in various medicines and rubbed on bodies. Finally, these forms of sexual license were apparently used as remuneration for services (such as healing) and were also expected offerings at feasts and dances.

Here, ideas about fertility, reproduction, health, illness, and exchange—and above all, about the remarkable powers of sperm—all produce a set of sexual practices that could not be more effective in confounding certainty of *individual* paternity. The whole point is that multiple acts of insemination are required for reproduction and, therefore, neither

certainty of individual paternity nor individual proprietariness (or jealousy) was a relevant category. The
final irony was that these particular forms of abundant
spermatic expenditure resulted in a low fertility rate—
the presumption being that sterility resulted from the
chronic irritation of the females' genitals—and, as a
consequence, the Marind-anim were known to kidnap
children from other groups to supplement their own
reproduction. It is hard to find an example that more
spectacularly confounds the universalist assumptions
of evolutionary psychologists.

One last example. Evolutionary psychologists
assert that the payment of bridewealth is, whatever
anthropologists might want to say about it, self-
evidently a sign of men's commodification of and
proprietary control over women. The assumption is
that bridewealth is a form of outright payment for
women, and that it establishes a relationship of owner-
ship over the woman. But evolutionary psychologists
fail to understand systems of gift exchange, which
explicitly challenge the logic of commodification. In
the Tanimbar Islands, and, I expect, many other
places, the marriage of a woman does not imply the
sale of a man's sister or daughter. Indeed, for a man to
give his sister in marriage is to give a part of himself, a
part that can never be fully alienated, and a man
remains the "owner" or "master" of the female lines of
sisters and daughters that emanate from his house. Far
from separating a woman from her house, the gift of
the woman indebts her husband and incorporates both
him and any potential children into her brother's
house. In Tanimbar, what bridewealth payments effect

is not the *sale* of the woman but rather the *redemption* of the man. In order not to remain permanently incorporated into his brother-in-law's house (or, worse, to be enslaved), the husband must "redeem" himself and his children through a series of prestations that establish the residence and affiliation of the man, his wife, and their children to the house of his own father. According to Tanimbarese understandings, a man remains permanently indebted to his own and his children's maternal relatives for fertility, life, and health, and this indebtedness is expressed in the structure of exchanges over the course of several generations. In Tanimbar and elsewhere, it is hard to conceptualize bridewealth as the sale of a woman when it is a sign of the indebtedness of the man, is the means of his redemption from the potential fate of slavery, is part and parcel of a larger system of exchange that extends across multiple generations, and has, as its purpose, the facilitation of life, health, and fertility. Because, for evolutionary psychologists, everything flows from the logic of individual interest, self-maximization, profit, and proprietary commodification, it is not at all surprising that they misconstrue the logic of systems of exchange, which organize the relationships between people, and between people and objects, in accordance with other principles of value.

From "Core Mindset" to Cultural Meaning

It is clear that even a cursory review of the anthropological record can not support evolutionary psychologists' claims to the universality of their most basic presuppositions. Although it may be true that females everywhere seek spouses with resources, it would be an historical and cross-cultural anomaly were the inverse not also true. Whereas the double standard and male proprietariness over females may exist in some cultures, there are others whose sexual practices and gender relations utterly contravene the presumption of their universality. What are we to make of this disjunction between the universal claims of evolutionary psychology and the diversity of human cultural ideas and practices? By way of concluding this ethnographic excursion, I wish to explore this disjunction to ask what it says about the theory of mind and culture proposed by evolutionary psychology; what it says about the ethnocentrism and naturalization their theory requires; and what it says about the nature of meaning in human cultures.

First, as noted earlier, to deal with cultural variation, evolutionary psychologists rely upon a distinction between genotype and phenotype. How effective is this in resolving the disjunction between their proposed universals and the evidence of cultural variation? Their argument is that the genotypical form—say, female preference for males with resources—is manifest when their presumed universal appears, whereas the phenotypical form—such as male preference for females with resources—is manifest when some culturally variant form appears. But, if evolutionary psychologists actually did attempt to take account of the full range of human social arrangements, they would end up having to posit some kind of frantic switching mechanism to deal with the discrepancies between the genotype they posit and the variable phenotypes that clearly exist in the world. A module that mandated male heterosexual promiscuity would have to be switched off in the highlands of New Guinea, in Thai monasteries, and in a host of other places. A module that failed to encode a male preference for women with resources would have to be switched off through most of human history. The "Madonna-whore switch" would have to be turned off in Lesu and the Trobriands, in polyandrous societies, and elsewhere. Or, to use a linguistic example, a psychological mechanism that supposedly codes for a universal distinction between nouns and verbs would have to be switched off in those human languages where such a distinction is absent or irrelevant.

Ultimately, such a plethora of switching mechanisms is hardly parsimonious. More importantly, to the extent that evolutionary psychologists account for social forms that diverge from their universals by an appeal to "cultural factors," they recognize the force of cultural creativity in at least some instances. But if, as they acknowledge, cultural factors operate in some cases, it is unclear why they should not operate in all cases, or even how it might be determined when they are or are not operating. It would be far more parsimonious to assume that the evident flexibility of the human brain makes it capable of creating a range of meaningful cultural orders—some of which happen to look like the models that evolutionary psychologists privilege as universal, and many of which happen to look quite different.

Second, the reduction of the variety of human cultural forms to a genotypical "core mind set" that looks suspiciously Euro-American in its valorization of the individual, of genetics, of utilitarian theories of self-maximization, and of a 1950s version of gender relations, has two effects. On the one hand, it naturalizes dominant Euro-American assumptions about sexuality, gender, and kinship. Indeed, by beginning with a deductive hypothesis that has already presupposed the ultimate causes of social behavior and already made universals out of their own cultural categories, they have completed the process of naturalization before they even begin. On the other hand, they need never put their theory at risk by confronting distinctly different cultural categories and, moreover, they effectively erase what we know about the complexity and diversity of human cultures around the world and across time.

Finally, if the cross-cultural diversity of cultural effects can not be reduced to the same common cause, or "core mindset," it is not simply because this core mindset happens to be, in the end, a culturally specific mindset masquerading as a universal and natural cause. It is also because culture provides the meaningful relationships—the ideas, beliefs, values, and practices—through which humans mediate the relation between cause and effect. The process of mediation goes both ways. For not only may the same cause have different effects, but the same effect may have different causes.

It is clear from the ethnographic materials discussed in this section that a particular psychological propensity—male jealousy, for example—can be manifested through a multitude of different cultural conventions, or not at all. Evolutionary psychologists would grant this, providing they could ultimately reduce the multiple cultural realizations or effects to the same causal "core mindset," with the attendant problems just noted.

But there is a more subtle form of cultural mediation that was articulated long ago by the founding father of American cultural anthropology, Franz Boas, who argued that, in the realm of culture, "like effects do not necessarily have like causes." This is because cultural effects that might appear similar have been constituted through very different sets of meanings. Take active female sexuality. Under Victorian sexual mores, the cultural valuation of a sexually active woman is constituted in a system of meanings that contrasts normative male promiscuity with female coyness, "loose" female sexuality with sexual reserve, and whores with the figure of the Madonna. In the !Kung, Pokot, Trobriand, Nage or Lesu cultural systems, a sexually active woman can not even be called promiscuous, since activity outside of marriage is the norm—for both sexes—and sexual experience is valued as a prelude to marriage. In the Nayar system, the active sexuality of women with multiple male partners can not be called promiscuous either, since it takes place within legally and ritually sanctioned polyandrous marriages. It is evident from these few examples that an effect that looks "objectively" the same—an active female sexuality—has very different causes and meanings, and therefore actually constitutes an array of quite different phenomena.

Anthropologists Stefan Helmreich and Heather Paxson have made a similar argument in response to the assumption, in Randy Thornhill and Craig Palmer's *A Natural History of Rape*, that rape is everywhere the precipitate of the same underlying logic of male genetic maximization. They argue that, however similar rapes may appear, they can not be understood to be the same phenomenon precipitated by the same underlying cause because, they have very different meanings in different contexts. They contend that, in the context of an ethnic nationalist war like that which took place in Bosnia-Herzegovina in 1992, where over "twenty thousand women reportedly were raped," the act "exceeds the physically sexual to become a highly orchestrated strategic instrument of war. It is not paternity that is being maximized here; it is a focused collective effort to terrorize, and destroy the cultural integrity of, the vanquished group." By contrast, in the context of slavery in the United States before the civil war, "rape was about property owner-ship and economic advantage, not an evolutionarily selected drive to ensure males' genetic contribution to the next generation." And, in the context of American university life, fraternity gang rape is a "form of male bonding" and "a rite of male camaraderie, not male competition."

Thus, where evolutionary psychologists posit universal causes and psychological mechanisms that produce what they see as the same effect (e.g., rape), cultural anthropologists see behaviors that look, at first glance, to be objectively the same as fundamen-tally different, because they are constituted in accordance with fundamentally different cultural meanings. Evolutionary psychologists will say that the "objective"—that is, genetic—effect is the same anyway, but in the process they wipe away the entire world of human cultural meaning and intention, and their consequences.

V. SCIENCE AND FICTION

The practice of science, like all human activity, depends upon categories, understandings, and conventions of practice that are, inevitably, culturally and historically specific. As noted earlier, the point is not that "good science" operates outside of culture and without reference to cultural categories, while "bad science" does not. On the contrary, it is precisely because "good science" recognizes its inevitable situatedness within culture that it must always place its most fundamental categories, understandings, and conventions at risk through the examination of contrary evidence. At least ideally, the scientific method requires that a hypothesis be tested against empirical data that have the potential for disproving it—that is, against aspects of the world that are relevant, resistant, and not already internally implicated in its own presuppositions. It is precisely evolutionary psychology's failure to do this that makes it "bad science."

Wright suggests that those who have argued against "innate mental differences between men and women... [have] depended on the lowest imaginable 'standards of evidence'—no real evidence whatsoever, not to mention the blatant and arrogant disregard of folk wisdom in every culture on the planet." But I would turn this accusation back on evolutionary psychologists, for they have constructed their theories through inappropriate analogies and data sets, limited and biased sampling, a host of unsubstantiated conjectures, and sheer fabrication. An examination of the ways in which evolutionary psychologists construct their "evidence" is lesson in how "bad science" comes into being.

To begin with, it is painfully evident from their writings that evolutionary psychologists know a lot about insects and birds but very little about humans. Daly and Wilson note that "within human evolutionary psychology, much of the best work is conducted by animal behaviourists who treat *H. sapiens* as 'just another animal.'" Indeed, while evolutionary psychologists may have conducted field research on kangaroo rats, they rarely conduct field research in human societies. While they may be experts on mating among gladiator frogs, they are extraordinarily ignorant of the extensive literature on the varieties of human gender, sexuality, kinship, and marriage. While they purport to know about the deep structure of all languages, they have rarely bothered to attain fluency in any non-western language. And, while they feel confident in attributing cultural properties to animals, they have rarely attempted to unravel the complexities of a single human culture. The question then becomes what, in this state of extreme ignorance, comes to count as knowledge about culture and cultural difference?

Organic and Cross-Species Analogies

Although evolutionary psychologists present themselves as "real" scientists, interested only in the cold hard facts, they use a wide range of literary devices to create what is perhaps a compelling fiction but one that hardly stands up to the evidentiary requirements of a hard science, or even a soft one. Two of the central rhetorical tools evolutionary psychologists use are the analogies they draw between social and organic processes, on the one hand, and between humans and other species, on the other.

Generally, one of the first lines of argument evolutionary psychologists mobilize in their accounts of human relations involves an analogy between those "preference mechanisms" relating to "mate selection" and those relating to organic processes, such as eating. Yet, predictably, they do not focus on the remarkable creativity of human food systems but rather on that aspect which is most automatic and unconscious in human responses to food—the gagging, spitting, or vomiting of foods deemed to be repulsive. Humans are presumed to choose certain mates and reject others just like they relish certain foods and have a gag response or aversion to others. Disregarding anthropological accounts of the diversity of cultural understandings of what counts as food—enticing, repulsive, or otherwise—let alone "mates," the rhetorical strategy here is to equate a manifestly social process (selection of a marriage partner) with what is assumed to be a manifestly organic process (avoidance of non-nutritious foods and selection of nutritious foods). Thus, evolutionary psychologists suggest that women "may" have the equivalent of a gag response to men without resources in the same way that humans have a gag response to putrid foods. Although the character of such assertions is entirely hypothetical (marked by the use of the conditional modality—"women *may* do X"), they have the effect of attributing natural, non-conscious, and automatic qualities to what is otherwise a social, highly conscious, and culturally mediated process. The analogy collapses, moreover, with even cursory attention to the cross-cultural variability of food systems. Whereas the putrid smells of some foods (for instance, durian or well ripened cheeses) may evoke a gag response in the uninitiated, they are the source of refined discrimination among the initiated in cultures where these foods are prized. The thought of eating dog, or pork, or beef will elicit repulsion differentially in Americans, Muslims, or Hindus. Rather than automatic and unconscious, taste in food is culturally variable and learned, the subject of extensive cultural commentary, and used as a signifier of ethnic, class, and religious difference.

Next in order of argumentation after the organic analogies are the cross-species analogies. In order to make such analogies make sense, evolutionary psychologists assert that many of the mental "mechanisms" relating to "mate choice" are not only *cross-cultural* universals but also *cross-species* universals—an assertion that seems to imply that there is a psychic unity of all species, not just all humans! Thus, evolutionary psychologists do not hesitate to draw direct analogies between the purported preferences and choices of a host of insects, birds, and mammals and those of humans. Take Buss' account of the weaverbird. The male weaverbird builds a nest and, in order to attract a female, suspends "himself upside down from the bottom and vigorously flap[s] his wings." If a female is interested, she comes by and inspects the nest, while the male sings to her. Note the choice of words and the immediacy of the narrative link drawn between birds and humans.

> At any point in this sequence she may *decide* that the nest does not meet her *standards* and depart to inspect another male's nest.... By *exerting a preference* for males who can build a superior nest, the female weaverbird *solves the problems* of protecting and provisioning her future chicks. Her *preferences* have evolved because they bestowed a reproductive advantage over other weaverbirds who had no preferences and who mated with any males who happened along.
>
> Women, like weaverbirds, prefer men with desirable "nests."

Such narratives involve two reciprocal analogical transpositions. First, animals are "like" humans and are given human qualities. Here, the weaverbird has human mental capabilities—the ability to *decide*, to *exert a preference*, and to *choose*—and operates in the context of a complex cultural hierarchy of values—she has "*standards*" and "*preferences*." Second, humans are "like" animals and are given animal qualities. Thus, a woman likes a good nest as much as a female weaverbird, or "[l]ike zebra finches, human affairs seem to be affected by the relative mate value of the partners." Evolutionary psychologists, like the zoopsychologists of a century ago, spin the web of their arguments with these analogic imaginaries. Scorpionflies (like men) select and give "substantial nuptial gift[s]" to attract their mates; male swallows (like male humans) copulate by force; ring doves (like humans) have a 25 percent divorce rate every season; female gray shrikes (like women) avoid "entirely males without resources, consigning them to bachelorhood"; dunnocks (like humans) can be found in "[m]onogamous pairs, polyandrous trios, polygynous trios, and even polygynandrous groups."

These analogies both assume and are meant to establish the existence of a universal and unconscious reproductive logic that underlies the behavior not only of all *humans* but of all *species*. Yet this claim of cross-species psychic unity is an artifact of a transposition of attributions: while human social institutions (marriage, divorce) and values (standards, preferences) are attributed to non-human species, terms for animal behavior are used to describe human behavior (marriage is consistently referred to as mating). In the process, the social institutions and values that are central to specifically human relations are entirely erased. As Eleanor Leacock suggests, such transposition violates a fundamental principle of evolutionary theory, since it "takes behavior from different phylogenetic levels which is merely analogous and derived from varying causes, and implies it to be homologous and derived from the same causes."

This claim of cross-species sameness is also the artifact of highly selective comparisons of animal and human protagonists. One could tell a story markedly different from that propagated by the evolutionary psychologists if one simply chose a different set of animal protagonists. One hardly needs to go so far as scorpionflies to make the point. Among our close primate relatives, two species of chimps will do. Biologist Anne Fausto-Sterling queries: "Which shall we choose as our model female? Females of the better-known chimp species have an associated pattern of hormones and copulation, but the bonobo female has sex constantly with both males and females and apparently uses sex not just for reproduction, but as a medium of social mediation." Attention to the polymorphous sexual behavior of bonobo females would clearly disrupt the stereotypic accounts of gender and kinship propagated by evolutionary psychologists, and they are therefore careful to avoid such examples. Instead they range widely amongst the species—from spotted sandpipers to gladiator frogs to elephant seals—to find animal behavior that will confirm the gender stereotypes that are central to their narrative.

In the end, the analogies between the lives of scorpionflies or elephant seals and humans are possible because the distinctive features of humans—large brains capable of inventing a wide variety of culturally distinctive behaviors—are entirely absent from the accounts of evolutionary psychologists. For evolutionary psychologists, decisions and choices *in all species* "reside within the organism," as innate, genetically transmitted, non-conscious processes driven by a presumptively universal logic of genetic self-interest. In the process, the human mental capacities of reason, emotion, and choice have been removed from consciousness and been thoroughly naturalized and geneticized, at the same time that the role of culture as a conceptual framework that mediates human experience of and behavior in the world has been erased. Paradoxically, the surfeit of analogies produced by evolutionary psychologists is a good example of the creativity of the human mind—its ability to improvise novel symbolic constructs—even as it is hardly evidence of serious scientific inquiry.

The Fabrication of Cross-cultural Deep Structures

The naturalization of psychological mechanisms into a geneticized deep structure allows evolutionary psychologists to posit cultural diversity as a kind of phenotypical surface structure that is triggered by differential "environmental" factors. While acknowledging cultural complexity and diversity, evolutionary psychologists give primacy, as we have seen, to what they argue is "the ubiquity of a core mindset." Much rests on the presumed universality of this "core mindset." The assertion of both the innate nature of these mechanisms and their origin in a primal evolutionary adaptive environment hang by a thread upon this universality.

Thus, in an effort to establish the universality of psychological mechanisms, evolutionary psychologists mobilize an extensive array of "preference" studies. These studies are generally marshaled in a particular narrative order: after the studies of insects, birds and bees, come those of United States college students, then Americans more generally, then hunter-gatherer societies and rare snippets from anthropological ethnographies, and finally Buss' often-referenced survey of 37 societies. From the wide range of human cultures that might be sampled, how representative and reliable are the samples offered in the studies of evolutionary psychologists?

The primary and most extensive data on "mating preferences" come from the United States, and not from a random sample of Americans but from the rather skewed population of Americans most readily accessible to university researchers: undergraduate students, aged 17-21. Occasionally, evolutionary psychologists will consult or conduct studies that reach other populations of Americans—such as those who write personal ads or those who frequent singles bars. But, by far, the most frequently surveyed group is that of university students—a "captive" group whose "paper-and-pencil responses," Wilson and Daly admit, "may or may not have anything to do with anything they have ever experienced." Although such a population is hardly representative of Americans, in general, it is very often made to stand as representative of the human species.

It is not that evolutionary psychologists are unaware of anthropological studies of other cultures. However, they are exceedingly selective about whom they read, what kinds of societies are worth reading about, and what details are worth noting. Because they assume that psychological mechanisms originated in an environment of early evolutionary adaptation, which they imagine as a hunter-gatherer society, they are most interested in what they see as the contemporary "relics" of the original hunter-gatherer societies. Moreover, they display a distinct preference for reading only those authors—for instance Napoleon Chagnon on the Yanomamö of Brazil, or Kim Hill and A. Magdalena Hurtado on the Ache of Paraguay—whose biologically reductive accounts most accord with their own preconceptions of what hunter-gatherer societies are all about. They fail to consult other authors—such as Ann Fienup-Riordan on the Yup'ik Eskimo, or Signe Howell on the Chewong of Malaysia, to name just two—whose accounts of hunter-gatherers would severely compromise their theories; and they are "puzzled" when hunter-gatherer groups such as the Australian aborigines do not conform to their imaginary models. Moreover, they fail to understand that contemporary hunter-gatherer peoples are hardly isolated "relics," but rather live in complex relations with rural and urban neighbors, have been subject to long colonial and missionary histories, are citizens of nation-states, and participants in the new global cultural and political economy. Their presumption that contemporary hunter-gatherer groups provide windows on life in some imaginary evolutionary adaptive environment millions of years ago is therefore deeply problematic.

Evolutionary psychologists do, on rare occasions, expand their horizons beyond hunter-gatherer groups and consult other anthropological works. However, when they do, they are exceedingly selective in what they learn from such accounts. For example, Buss quotes Bronislaw Malinowski to the effect that Trobriand women will not offer sexual favors to men who do not display resources in the form of a gift, a detail he employs to support his thesis that women universally favor men with resources. However, he ignores the many details in Malinowski's work—including extensive evidence noted earlier of Trobriand premarital sexuality and women's sexual precociousness—that would seriously compromise evolutionary psychology's presuppositions about women's innate psychological mechanisms. Similarly, Daly and Wilson note that "in the Pacific island... of Tikopia, a man who acquired a wife who was already a mother would be forthcoming about his unwillingness to invest in a predecessor's child... demanding that the child be either fostered out or destroyed." They do not investigate, however, the extensive literature on the widespread incidence of adoption and fosterage in Oceania and elsewhere, since it does not support their argument that kinship can be reduced to an investment logic of genetic self-maximization.

Not only do evolutionary psychologists pick and choose favorable bits from the ethnographic record, they often outright misrepresent it. One example is typical. Daly, Wilson and Weghorst, in their review of the ethnographic literature on societies that have been reported to lack the double standard and male sexual jealousy, use Hortense Powdermaker's account of Lesu sexuality, which we examined earlier. The three authors acknowledge Powdermaker's report of Lesu women's extramarital sexual relations and of their husbands' willingness to accept all the women's children as their own. However, they go on to discount the significance of Powdermaker's evidence, which runs contrary to their own presuppositions, by misrepresenting three details—two that relate to certainty of paternity and one that relates to jealousy. First, they suggest that Lesu "wives are concerned to avoid extramarital impregnation." One would assume, from reading this, that they mean that Lesu wives use some form of birth control in their extramarital sexual encounters. But what Powdermaker actually says is that "after they have had intercourse with their lover they have it the next day with their husband, and this makes all of the children his." Presumably, however, women's subsequent intercourse with their husbands secures social, but not always genetic paternity—that is, unless one believes that "sperm competition" always protects the paternity of legal husbands over extramarital lovers. Lesu women actually do use a form of birth control, not specifically to avoid extramarital pregnancy, Powdermaker says, but rather to avoid the pain of birth again and to ensure that they will be able to continue to engage in ritual dances and further sexual adventures.

Second, Daly, Wilson, and Weghorst note that if an unmarried woman gets pregnant, her lover is not obliged to marry her, and they go on to quote the first of Powdermaker's two reasons ("he does not know if the child is really his, as the woman has probably had several lovers") but not her second reason ("all he wanted was intercourse, not marriage, and the Melanesian distinguishes very clearly between the two"). Daly, Wilson, and Weghorst do not appear to understand how to integrate the second reason with the earlier noted fact about husbands accepting all of their wives children as their own. The point is that it is *marriage* that carries responsibilities toward a woman's children, not *sexual intercourse*. Children born outside of marriage are not stigmatized, and no woman expects premarital or extramarital partners to take responsibility for her children: this is the responsibility of the man she marries, and he accepts this responsibility willingly.

Third, Daly, Wilson, and Weghorst argue that "Powdermaker describes wife-beating as a punishment for adultery in both case histories... and folk-tales...." They fail, however, to relate many details of Powdermaker's account: that jealousy over extramarital relations is by far the exception, not the norm; that, on the rare occasions that jealousy is expressed, it is as likely to be women as men who express it (two of the four cases of jealousy she reports involve women); that women as well as men are capable of expressing violent jealousy; that in one of the wife-beating cases they refer to, the husband beats his wife because she is expressing *her* jealousy of *his* flirtations with women, not the reverse; that this same man—one of only two men reported to have expressed violent jealousy—was psychologically unstable and prone to unconscious "mad" fits, which had periodically recurred since childhood; and, finally, that, in addition to the folk-tales that feature instances of jealousy, there "are many other folk-tales in which men have mistresses and wives lovers, but in which there is no jealousy." When all the details of Powdermaker's account are reviewed, it is hard to see how the Lesu material could be mobilized as evidence for their claims for a universal double standard and violent male sexual jealousy, or how Daly, Wilson, and Weghorst could dismiss its ample evidence to the contrary, which they do.

My point here is not to dicker with the details. Rather, I wish to indicate the kind of scholarship that is at work in these accounts. Not only do evolutionary psychologists pick and chose the details that fit with their preconceptions—and often misread these details at that—but they also simply ignore the complex cultural whole from which these details were selected as well as the larger anthropological literature on relevant topics. In short, evolutionary psychologists shape "evidence" to fit preconceived notions—a practice that too often characterizes deductive reasoning in the social sciences.

Because the large library of anthropological descriptions of other systems of social life is clearly not supportive of the aims and claims of evolutionary psychology, Buss and his colleagues have conducted their own survey of "human mate preferences" in 37 cultures. One might excuse the dismissal of anthropological literature and the heavy reliance on this particular study, if the study were, in any way, a reliable account of cultural differences. However, many people—including Buss himself—point out the considerable shortcomings of this study. Not only are 27 of the 37 societies in the sample either European or strongly influenced by European culture, but the sample is also "biased toward urbanized, cash-economy cultures" over rural, non-cash economies. Moreover, the very nature of the survey "instrument" includes *a priori* categories, many of which derive from theory-driven assumptions as to what criteria are relevant—for instance, good financial prospect, favorable social status, chastity, good looks. At the same time, it excludes a number of categories that any anthropologist would assume to be relevant to a cross-cultural study of marriage—for instance, cousin relations, hypergamy and hypogamy, exogamy and endogamy, caste, race, and religion, to name a few. Finally, it neglects utterly to elicit categories that respondents themselves might deem relevant.

The considerable limitations of the study aside, its results are most remarkable in their failure to support the predicted gender-differentiated preference mechanisms. Of the eighteen possible characteristics that the survey investigates, the first four (mutual attraction, dependable character, emotional stability and maturity, and pleasing disposition) are ranked in the same order by both males and females, and the next four (good health, education and intelligence, sociability, and desire for home and children) include the same categories for both males and females although not necessarily in the same order. None of these criteria supports the proposition that mate preference is guided by gender-differentiated preference mechanisms of the sort outlined by evolutionary psychologists. In fact, as Buss admits, these data suggest that people of both genders place a number of criteria far above those central to the evolutionary thesis, "suggesting that species-typical mate preferences may be more potent than sex-linked preferences."

Not only do the eight top-ranked criteria show no appreciable link to the evolutionary thesis of gender differentiated preferences, but two of the five predictions—those regarding female preferences of ambition and industriousness in men, and male preferences of chastity in women—showed high cultural variation. As noted earlier, the significance of cultural variation that contradicts the presumed universals of evolutionary psychologists is either outright dismissed or accounted for by a sudden appeal to "cultural factors" that otherwise are granted no place in the narratives of evolutionary psychology. Yet, evolutionary psychologists never specify when or how cultural or "environmental" factors come into play, what their relationship is to the genetically determined psychological mechanisms, or why they operate sporadically.

Finally, there is a pervasive ethnocentrism in the accounts of evolutionary psychologists. What is assumed to be universal and innate is, in the end, simply the researcher's own cultural categories in disguise. Cultural formations that do not accord with their own cultural presuppositions are deemed to be secondary responses—phenotypical not genotypical. The cultural and historical specificity of these supposedly universal and innate categories is invisible to the researchers because they have already completed the task of naturalizing them before they have begun to investigate them cross-culturally.

Evolutionary and Genetic History: The Cartoon Version

Even if the evolutionary psychologists could find a universal "psychological preference mechanism," there would still be two crucial steps left in the argument. They would have to show not only that it originated in the Pleistocene environment of evolutionary adaptation in order to serve some adaptive purpose but also that it has been transmitted genetically from the Pleistocene to the present. It is, however, at this point in the argument that pure conjecture—which they call reverse-engineering—takes over and can be recognized, grammatically, by the shift into a conditional modality.

Among the many problems with their conjectures about evolutionary origins, two stand out: a remarkable vagueness about the time, place, and circumstances of the environment of evolutionary adaptation; and an equally remarkable specificity about the nature of social relations in that hypothetical time and place. Given how much importance is placed on this original "environment of evolutionary adaptation," it is perplexing that its specific characteristics are rarely described or analyzed. But, as Fausto-Sterling argues,

> It is not unreasonable to ask the hypothesis-builders of evolutionary psychology at least to postulate at what point in human or hominid history they imagine contemporary reproductive behaviors to have first appeared. "Throughout the Pleistocene" is pretty vague. What is the evidence that it wasn't earlier or later?... What were the food and predator stresses at that moment? Data on these points can be gleaned from the archaeological and geological record. How did humans respond? Biogeographic data can be brought to bear on this point. Was there a division of labor during this early period of evolution? Or did gender-based divisions of labor evolve later?

Evolutionary psychologists appear to be uninterested in these and many other questions about the nature of the environment of evolutionary adaptation. They also appear to be ignorant of—or uninterested in—the relevant paleontological, archaeological, geological, and climatological evidence and unwilling to test their theories in light of this evidence.

What would the significance be, for their theories, of the evidence that there was not one single and unchanging environment of evolutionary adaptation? Drawing upon the work of Richard Potts, the Director of the Human Origins Program in the Smithsonian's National Museum of Natural History, Kathleen

Gibson cites data showing that hominid adaptive environments were variable and fluctuating.

> By about 2.4 million years ago, savanna habitats were expanding, and by about 1.8 million years ago, fully bipedal hominids had appeared. During this period, however, and throughout the subsequent Pleistocene, evidence indicates that mean global temperatures and sea levels fluctuated frequently, leading to periodic changes in terrestrial climates. For example, at one fossil site, Olduvai Gorge, the habitat was at times a relatively moist, humid, lakeside environment and at other times dry and semi-arid. These climatic fluctuations resulted in repeated, major changes in the fauna and flora available for human consumption.

Evolutionary psychologists recognize that their concept, the "environment of evolutionary adaptation," is "a fiction, a composite drawing," and that the ancestral environment actually "changed much in the course of human evolution." But, in face of the fact that evolutionary adaptations always happen in relation to *local* environments, not to generalized fictional ones, the fiction is actually meaningless and misleading.

Moreover, the continued evocation of a single, unchanging fictional environment allows evolutionary psychologists to suppress the significance of the variability of changing ancestral environments, since it would threaten their theory of mind and culture. For, as Gibson points out, the fluctuating and variable nature of the ancestral environments—and their flora and fauna—indicate that prior to the emergence of our species, hominid predecessors had demonstrated a behavioral versatility that enabled them to survive in a diversity of climatic and geographical conditions. This suggests that natural selection favored those hominids with the neural and mental capacities to solve novel problems rather than those able to solve only those problems encountered by their ancestors.

Yet evolutionary psychologists do not test their theories against this more realistic portrait of the ancestral environments of humans. Content to operate with a "fiction," we are left to conjure up pictures of life in the Pleistocene from our memories of natural history museum dioramas and 1950s TV family dramas.

While the specific characteristics of the original adaptive environment are entirely lacking in the accounts of evolutionary psychologists, the characteristics of social relations in that environment are given in stunning specificity, despite the fact that our ability to know anything about these relations is negligible. Paleoanthropologists are able to read certain things from the bones and stones that constitute the available evidence—for instance, about physiology, diet, and food production. But it is impossible to read the specifics of social relations (including those of sex, gender, kinship, marriage) from the available fossil evidence.

If social organization is impossible to surmise from the fossil record, how can we read psychology from fragments of bones and stones? It is precisely because the fossil record is and will forever be entirely mute on psychology that the evolutionary psychologists can (and must, if they are to have a story to tell) fill the void with the details of their own making. These inevitably end up being contemporary (if not Victorian) stereotypes of gendered psychology projected back into deep evolutionary history under the banner of "reverse engineering." But, as Hilary Rose notes, "[i]f Lucy's sex is a matter for technical debate, claims of certainty about prehistoric innate psyches look extraordinarily thin."

Yet we are asked not only to take the cartoon representations of our ancestors as serious representations of our evolutionary origins but also to believe that nothing has changed in the intervening millennia. Despite the creation and demise of cultural worlds of tremendous complexity and variety, despite everything else that is learned and changeable in the diverse human cultural landscape, we are asked to believe that human desires, motivations, and intentions got fixed once and for all and remain genetically programmed.

The same conjectural mode marks the jump from assertions of universals to assertions of innate inherited capacities. In a typical passage, Buss argues that "[s]exual selection *could have* directly fashioned, over the course of thousands of generations, psychological mechanisms that produce effective mate competition tactics," or again, that "contemporary men prefer young women because they have inherited from their male ancestors a preference that focused intently upon this cue to a woman's reproductive value." Here, the inheritance of unchanging psychological mechanisms is simply asserted, because it fits the theory and it fills the void of what we can never know about the social and psychological state of ancestral humans throughout the vast expanse of human prehistory. In the end, the most crucial steps in the evolutionary psychology argument—linking presumptively universal contemporary behavior to an ancestral evolutionary adaptive environment and to innate psychological mechanisms forged by the forces of natural and sexual selection—remain simply conjectures in a narrative of mythic origins.

Again, evolutionary psychologists are unwilling to allow what we *know* about human evolution to impinge on what they wish to *conjecture* about human evolution, or to deal with the consequence of that evidence for their theory of mind and culture. Take, for instance, the well-established fact that the development of culture did not follow but rather preceded and co-occurred with the biological emergence of *Homo sapiens*. As Clifford Geertz notes, "the transition to the cultural mode of life took the genus *Homo* several million years to accomplish; and stretched out in such a manner, it involved not one or a handful of marginal genetic changes but a long, complex, and closely ordered sequence of them." Thus, the development of *cultural* adaptations was critical to the development of *biological* adaptations. As a consequence of this, Sahlins adds,

> [i]t is reasonable to suppose that the dispositions we observe in modern man, and notably the capacity—indeed, the necessity—to organize and define these dispositions symbolically, are effects of a prolonged cultural selection.... When the full implications of

this simple but powerful argument are finally drawn,
a great deal of what passes today for the biological
"basis" of human behavior will be better understood
as the cultural mediation of the organism.

Evolutionary psychologists' insistence on the innate
nature of human behavior and psychology thus
suppresses not only the role of culture in the process of
their development but also its role in human biological
evolution. A consideration of what we know about the
significance of the human mental capacity for culture as
an active factor in human evolutionary history would
seriously challenge a theory that understands the
human mind as a passive enactor of natural selection's
designs, and human culture as an epiphenomenon of
genetic competition.

Ultimately, the "scientific" account provided by
evolutionary psychology is a fiction created out of
specious cross-species analogies, a plethora of undergraduate surveys, a paucity of research experience with
actual human languages and cultures, a refusal to examine seriously the cross-cultural, historical, or paleoar-
chaeological records, the summary dismissal of other
explanations and evidence, a host of fantasy genes, and
a fairy tale about evolutionary origins.

It is not surprising that, given the absence of
reliable and appropriate evidence to support their
account of human behavior, some commentators have
noted that the "science" of evolutionary psychology
looks more like a form of religious fundamentalism, in
which we are asked to accept an authoritative version of
reality on faith alone.

VI. SCIENCE AND MORALITY

Steven Pinker has charged that the critics of evolutionary psychology wish to view the world through
rose-tinted glasses that blind them to the darker side
of the human condition. Evolutionary psychologists
argue that, however tough it is to acknowledge the
darker side of that nature, someone has to do it, and
their job is to shed the cold light of scientific realism
on human nature, including its more unsavory bits.
They stress that they—unlike other social scientists—
have the key to the fundamental realities of "human
nature." It is only by understanding the evolutionary
logic of genetic maximization and natural selection,
and the innate psychological mechanisms that define

human nature that we can deal socially with their consequences. The "disturbing side of human mating [jealousy, rape, incest, violence, etc.] must be confronted," Buss argues, "if its harsh consequences are ever to be ameliorated." But their cold light of scientific realism has been conjured, as we have seen, out of a decidedly unscientific methodology and with a congenital blindness to the significance of human cultural variation.

Despite the flimsiness of their argumentation, evolutionary psychology is compelling because it gathers into one grand narrative a number of beliefs that are central to Euro-American culture—about the innateness of gender differences and of the sexual double standard; about the naturalness of neo-liberal economic values of self-interest, competition, rational choice, and the power of the market to create social relations; about the survival of the fittest and the determinant force of genes; about evolutionary origins and man the hunter; and about life's complexity having a single key. By combining such popular cultural beliefs into a gripping narrative and cloaking it in the guise of science, evolutionary psychologists give what are otherwise culturally specific "truths" the aura of a single, fundamental and universal truth.

Although the idea that the complexities of a particular phenomenon can be unlocked by a single, universal key might be a productive form of reductionism in physics, the poverty of such explanations of human social life has been consistently noted. Indeed, this kind of scientific reductionism has several serious consequences for how we might begin to think about contemporary social issues, let alone resolve them. It limits the kinds of questions we can ask about human social life and therefore the kinds of answers we might discover. It requires a studied ignorance of the varieties of human cultural forms or a steadfast dismissal of contrary evidence. It makes us as blind to alternative futures as we are to the diversity of historical pasts. And it is covertly prescriptive as it naturalizes a particular cultural representation and asserts its cross-cultural universality and inevitability.

Consider how the appeal to the universal explanatory power of natural selection and its presumed psychological mechanisms limits the kinds of questions one might ask and the kinds of explanations one might find compelling. Take Buss' argument that, however irrational male jealousy (including morbid jealousy) might seem, it is actually a rational evolved adaptation that arises specifically in response to the inevitable uncertainty of paternity. By narrowing the field of possible explanations of human behavior to the logic of genetic proliferation and to supposedly fixed psychological mechanisms originating in the Pleistocene, he and other evolutionary psychologists close off more obvious avenues of inquiry about the psychological origins and the varied social forms and dynamics of jealousy. It makes it impossible, moreover, to ask certain questions. Do females get jealous and why? Why does male or female jealousy manifest itself intensely in some cultures but hardly at all in others? Or, within any specific culture, why does it manifest itself in some people but not others? Because their question about the origin and function of jealousy contains critical presuppositions, it also already contains its own answers.

Consider how the naturalization of human culture and agency, and the corresponding dismissal of cultural variation, shapes the stories one can tell about the past and the questions one can ask about the present. It requires a studied suppression of the historical and ethnographic record to persist in the assumption that men universally control resources and that, in the environment of evolutionary adaptation, women (but not men) would have evolved a preference for a mate with resources. An examination of the evidence would require a different, more complex conclusion and provoke a host of further questions that are impossible to conceive under the reigning presupposition. How are the productive labors of men and women actually delineated throughout history and across cultures? How do these productive labors articulate with those of reproduction? What are the possible relations between gendered productive labor and the control over resources, and how do these relate to the origins and historical transformations of structures of hierarchy and power?

Consider, too, how the naturalization of human culture and agency makes it as impossible to imagine other futures as it is to imagine other pasts. To continue on the theme of male control over resources, Buss asserts: the "forces that originally caused the resource inequality between the sexes, namely women's preferences and men's competitive strategies, are the same forces that contribute to maintaining resource inequality today." The implication is that these forces have produced the same structures for millennia and are hardly likely to change. In face of such a fundamental and enduring "reality," it becomes difficult to imagine that other structures of relations might be within the scope of human *possibilities* (or even to recognize that, historically, they have been within the scope of human *actualities*). Thus, despite the frequent exhortations of evolutionary psychologists to the contrary, they see their task not so much as one in which we might take responsibility for "ameliorating" gender inequities, but rather—as the evolutionary kinship therapists would have it—as one in which we bring our contemporary gendered characteristics back into synch with those postulated attributes of our prehistoric savannah sisters and brothers. Buss outlines the program: "[f]ulfilling each other's evolved desires is the key to harmony between a man and woman.... Our evolved desires, in short, provide the essential ingredients for solving the mystery of harmony between the sexes." Wright concurs when he says that "therapists will be better equipped to make people happy once they understand what natural selection *does* 'want,' and how, with humans, it 'tries' to get it."

Given the reductionist logic of evolutionary psychology, it is inevitable that imagined futures reflect and reiterate imagined pasts. This is, in part, the message of the Posners' futuristic imaginaries about cloning. Even if the absence of sperm banks in the Pleistocene, according to their reckoning, allows us to explore the potential of the new reproductive technologies without innate aversions, the weight of the genetic logic of natural selection makes it inevitable that we shape our cloning futures according to the same logic that supposedly shaped our reproductive prehistory. Never mind that a host of ethnographers have documented wonderfully creative and culturally variant responses to the new reproductive technologies, evolutionary psychologists can only imagine a singular prehistoric logic that would shape the practice and use of cloning.

Or, consider how the naturalization of human culture and agency in the designs of natural selection makes it impossible to comprehend the origins or resolution of current global conflict in a meaningful way. A Charlottesville, Virginia weekly newspaper, *The Hook*, published an article called "Why They Kill: Male Bonding + Religion = Disaster," which was illustrated with photographs of "insurgent" Iraqis wrapped in headscarves and carrying rocket propelled grenade launchers. In the article, University of Virginia evolutionary psychologist J. Anderson Thomson argues that "[i]f we want to understand the genesis of terrorism—including September 11—we must confront the horrors of our evolutionary history, the murderous legacy it has left in men, and the violence that lurks at the core of religion." For the evolutionary part of his argument, he explains that adaptations

> always deal with fitness, the ability to pass genes to later generations. As man evolved, lethal raiding permitted males to successfully attract or secure reproductive-age females, weaken neighbors who

might compete for those women, inspire fear, protect themselves from incursion, expand their safe borders, and—by attacking in groups—incur very little risk.

Thomson weds his evolutionary argument about "male-bonded coalitionary violence" to one about religious intolerance to answer the question about the origin and dynamics of global terrorism. The irresponsibility of such an analysis and the ignorance it demands of us are monstrous. To understand the ongoing violence in Iraq and elsewhere, it is apparently unnecessary to know anything about the history or contemporary configurations of the relationship between Christianity, Judaism, and Islam, about the history of colonialism in the Middle East, about world imbalances of power and wealth, about the contemporary global political economy and neo-liberal capitalism, about oil reserves and oil markets, about western and Islamic ideas about nation-states and modernization, about U.S. imperial inclinations, or about the peculiar histories of American violence. It is enough, apparently, to know only that men's innate desire to proliferate their genetic legacy propels them to form male-bonded coalitions, and that when these are linked to religion, the result is "disaster." As Sahlins has argued, "a theory ought to be judged as much by the ignorance it demands as by the knowledge it purports to afford."

Finally, consider the moralizing effect of the fact that evolutionary psychology's "evolved desires" reflect the remarkable conjuncture of Victorian sexual norms and neo-liberal economic values that has come to dominate the current political scene in the United States. By naturalizing these norms and values into the structures of genes and the dynamics of deep evolutionary history, evolutionary psychologists produce what is, in effect, a set of moral prescriptions. Evolutionary psychologists claim to separate fact and value and they disclaim the prescriptiveness of their assertions. Thus Wright, for instance, declares that "[i]f... I seemed to suggest that women practice sexual restraint, the advice wasn't meant to carry any overtone of obligation. It was self-help, not moral philosophy." "Traditional morality" just happens to be in tune with the logic of natural selection: it "often embodies a certain utilitarian wisdom" and "is laden with practical, life-enhancing wisdom."

Yet, to the extent that certain cultural values are deemed to be true facts and universal adaptations, while others are dismissed as either unadaptive or irrelevant, their "facts" end up being deeply prescriptive and moralistic. Such un-self-reflective biases have enormous social and political consequences as certain people's understandings and practices are validated by the authority of science and made to appear natural and inevitable, while others are invalidated and made to appear unnatural, contingent, and superfluous. The claim is that evolutionary psychology provides a scientifically dispassionate, socially neutral view of humanity. The reality is that it is shaped by and serves to validate particular historical and cultural ideas, values, and practices that have—whether intentionally or not—specific political and economic affinities and alignments within the United States and globally.

Evolutionary psychologists attempt to escape the prescriptiveness of their naturalizations by suggesting that understanding the dark side of our inherited nature allows us to condemn it and ameliorate its consequences. As Buss advises, "Knowledge of our evolved sexual strategies gives us tremendous power to better our own lives by choosing actions and contexts that activate some strategies and deactivate others."

But, whether or not one agrees with their portrait of human nature, there is a final irony in this position. All of a sudden, humans are capable of conscious choice; all of a sudden they are able to create alternative cultural worlds when they so wish. Ultimately, then, in order to restore a redemptive element to their otherwise reductive and determinist project, evolutionary psychologists are forced to call upon the very qualities of human agency and creativity that their own theories essentially discount.

****

If culture consists, as the British anthropologist Marilyn Strathern and others have suggested, "in the way analogies are drawn between things, in the way certain thoughts are used to think others," then the web of analogies with which evolutionary psychologists have spun their myths and moral tales is excellent proof of the creativity, inventiveness, and arbitrariness of cultural constructions. Yet, it makes a difference what analogies we use to spin the web within which we suspend our lives as social beings. The difference is this: if culture is understood as the sum of individual, self-maximizing choices, and if those choices are naturalized into a geneticized firmament of evolutionary origins, then paradoxically this narrative ruled by the sign of choice tells a story about a lack of choice, about a world in which social hierarchies are fixed, human creativity annihilated, and only certain human realities are "true." Such a narrative not only erases the tremendous historical and contemporary evidence of human creativity and cultural diversity in the world—and the truths of other cultural realities—but it also severely constricts the kinds of questions we can ask and the kinds of social worlds we can possibly imagine and endeavor to create for ourselves. ■

Acknowledgements

I thank Marshall Sahlins and Matthew Engelke for the opportunity to contribute a volume to the Prickly Paradigm Pamphlet series, of which I am a great fan. Critical readings of manuscript drafts by Marshall Sahlins, Matthew Engelke, Amy Ninetto, and Joseph Hellweg improved the final version immensely. I am deeply grateful for their time, keen intellectual engagement with the project, and enthusiastic support. I would also like to express my appreciation to the Virginia Foundation for the Humanities, in whose cool, quiet offices I had the pleasure of spending the summer of 2005 and where I completed the revisions of *Neo-liberal Genetics*. Earlier versions of certain parts of this pamphlet were published in McKinnon 2002 and 2005.

Bibliography

Detailed notes for this title are available on Prickly Paradigm's website
as a free PDF file: www.prickly-paradigm.com/authors/mckinnon.html

Baal, J. van
    1966. *Dema: Description and Analysis of Marind-Anim Culture
    (South New Guinea)*. The Hague: Martinus Nijhoff.

Bailey, Kent G., and Helen E. Wood
    1998. Evolutionary Kinship Therapy: Basic Principles and
    Treatment Implications. *British Journal of Medical Psychology*
    71:509-523.

Barker-Benfield, Ben
    1972. The Spermatic Economy: A Nineteenth Century View of
    Sexuality. *Feminist Studies* 1:45-74.

Benton, Ted
    2000. Social Causes and Natural Relations. In *Alas, Poor Darwin:
    Arguments against Evolutionary Psychology*, edited by Hilary Rose
    and Steven Rose, pp. 249-272. New York: Harmony Books.

Birkhead, Tim R.
    2000a. Hidden Choices of Females. *Natural History* 11:66-71.

    2000b. *Promiscuity: An Evolutionary History of Sperm Competition*.
    Cambridge: Harvard University Press.

Boas, Franz
    1996 [1911]. Introduction to *Handbook of American Indian
    Languages*. Lincoln: University of Nebraska Press.

    1940. *Race, Language and Culture*. New York: The Free Press.

Bodenhorn, Barbara
    2000. "He Used to be My Relative": Exploring the Bases of
    Relatedness among Iñupiat of Northern Alaska. In *Cultures of
    Relatedness: New Approaches to the Study of Kinship*, edited by Janet
    Carsten, pp. 128-148. Cambridge: Cambridge University Press.

Borgia, Gerald
    1989. Typology and Human Mating Preferences. *Behavioral and
    Brain Sciences* 12:16-17.

Bowie, Fiona
    2004. *Cross-Cultural Approaches to Adoption*. London: Routledge.

Buss, David M.
    1988. Love Acts: The Evolutionary Biology of Love. In *The Psychology of Love*, edited by Robert J. Sternberg and Michael L. Barnes, pp. 100-118. New Haven: Yale University Press.

    1989. Sex Differences in Human Mate Preferences: Evolutionary Hypotheses Tested in 37 Cultures. *Behavioral and Brain Sciences* 12:1-49.

    1991. Evolutionary Personality Psychology. *Annual Review of Psychology* 42:459-491.

    1992. Mate Preference Mechanisms: Consequences for Partner Choice and Intrasexual Competition. In *The Adapted Mind: Evolutionary Psychology and the Generation of Culture*, edited by Jerome H. Barkow, Leda Cosmides, and John Tooby, pp. 249-266. New York: Oxford University Press.

    1994. *The Evolution of Desire: Strategies of Human Mating*. New York: Basic Books.

    2000. *The Dangerous Passion: Why Jealousy is as Necessary as Love and Sex*. New York: The Free Press.

Buss, David M., and David P. Schmidt
    1993. Sexual Strategies Theory: An Evolutionary Perspective on Human Mating. *Psychological Review* 100(2):204-232.

Buss, David M., et al.
    1990. International Preferences in Selecting Mates: A Study of 37 Cultures. *Journal of Cross-Cultural Psychology* 21(4):5-47.

Carroll, Vern, editor
    1970. *Adoption in Eastern Oceania*. Honolulu: University of Hawaii Press.

Carsten, Janet
    1997. *The Heat of the Hearth: The Process of Kinship in a Malay Fishing Community*. Oxford: Oxford University Press.

2001. Substantivism, Antisubstantivism, and Anti-antisubstantivism. In *Relative Values: Reconfiguring Kinship Studies*, edited by Sarah Franklin and Susan McKinnon, pp. 29-53. Durham: Duke University Press.

2004. *After Kinship*. Cambridge: Cambridge University Press.

Carsten, Janet, editor
2000. *Cultures of Relatedness: New Approaches to the Study of Kinship*. Cambridge: Cambridge University Press.

Carsten, Janet, and Stephen Hugh-Jones, editors
1995. *About the House: Lévi-Strauss and Beyond*. Cambridge: Cambridge University Press.

Chagnon, Napoleon
1968. *Yanomamö: The Fierce People*. New York: Holt, Rinehart and Winston.

Collier, Jane, and Sylvia Yanagisako, editors
1987. *Gender and Kinship: Essays Toward a Unified Analysis*. Stanford: Stanford University Press.

Coontz, Stephanie
1992. *The Way We Never Were: American Families and the Nostalgia Trap*. New York: Basic Books.

Daly, Martin, and Margo Wilson
1998. *The Truth about Cinderella: A Darwinian View of Parental Love*. New Haven: Yale University Press.

1999. Human Evolutionary Psychology and Animal Behaviour. *Animal Behaviour* 57:509-519.

Daly, Martin, Margo Wilson, and Suzanne J. Weghorst
1982. Male Sexual Jealousy. *Ethology and Sociobiology* 3:11-27.

Demos, John
1986. *Past, Present, Personal: The Family and the Life Course in American History*. New York: Oxford University Press.

Dolgin, Janet L.
1997. *Defining the Family: Law, Technology and Reproduction in an Uneasy Age*. New York: New York University Press.

Dover, Gabriel
2000. Anti-Dawkins. In *Alas, Poor Darwin: Arguments against Evolutionary Psychology*, edited by Hilary Rose and Steven Rose, pp. 55-77. New York: Harmony Books.

Draper, Patricia
1975. !Kung Women: Contrasts in Sexual Egalitarianism in Foraging and Sedentary Contexts. In *Toward an Anthropology of Women*, edited by Rayna R. Reiter, pp. 77-109. New York: Monthly Review Press.

Eagly, Alice H., and Wende Wood
1999. The Origins of Sex Differences in Human Behavior: Evolved Dispositions Versus Social Roles. *American Psychologist* 54(6):408-423.

Edgerton, Robert B.
1964. Pokot Intersexuality: An East African Example of the Resolution of Sexual Incongruity. *American Anthropologist* 66:1288-1299.

Evans-Pritchard, Edward Evan
1940. *The Nuer: A Description of the Modes of Livelihood and Political Institutions of a Nilotic People*. Oxford: Oxford University Press.

1951. *Kinship and Marriage among the Nuer*. Oxford: Oxford University Press.

Fausto-Sterling, Anne
1993. *Myths of Gender: Biological Theories about Women and Men*. New York: Basic Books.

2000a. Beyond Difference: Feminism and Evolutionary Psychology. In *Alas, Poor Darwin: Arguments against Evolutionary Psychology*, edited by Hilary Rose and Steven Rose, pp. 209-228. New York: Harmony Books.

2000b. *Sexing the Body: Gender Politics and the Construction of Sexuality*. New York: Basic Books.

Fienup-Riordan, Ann
1983. *The Nelson Island Eskimo: Social Structure and Ritual Distribution*. Anchorage: Alaska Pacific University Press.

1990. *Eskimo Essays: Yup'ik Lives and How We See Them*. New Brunswick: Rutgers University Press.

Fienup-Riordan, Ann, with William Tyson, Paul John, Marie Meade, and John Active
2000. *Hunting Tradition in a Changing World: Yup'ik Lives in Alaska Today*. New Brunswick: Rutgers University Press.

Foley, William A.
1997. *Anthropological Linguistics: An Introduction*. Cambridge, MA: Blackwell.

2005. Do Humans Have Innate Mental Structures? Some Arguments from Linguistics. In *Complexities: Beyond Nature and Nurture*, edited by Susan McKinnon and Sydel Silverman, pp. 43-63. Chicago: Chicago University Press.

Forth, Gregory
2004. Public Affairs: Institutionalized Nonmarital Sex in an Eastern Indonesian Society. *Bijdragen tot de Taal-, Land- en Volkenkunde* 160(2/3):315-338.

Fraga, Mario F, et al.
2005. Epigenetic Differences Arise During the Lifetime of Monozygotic Twins. *Proceedings of the National Academy of Sciences* 102:10604-10609 (published online before print as 10.1073/pnas.0500398102).

Franklin, Sarah
1997. *Embodied Progress: A Cultural Account of Assisted Conception*. New York: Routledge.

Franklin, Sarah, and Susan McKinnon, editors
2001. *Relative Values: Reconfiguring Kinship Studies*. Durham: Duke University Press.

Friedl, Ernestine
1975. *Women and Men: An Anthropologist's View*. New York: Holt, Rinehart, and Winston.

Geertz, Clifford
1973. *The Interpretation of Cultures*. New York: Basic Books.

Gibson, Kathleen R.
2005. Epigenesis, Brain Plasticity, and Behavioral Versatility: Alternatives to Standard Evolutionary Psychology Models. In *Complexities: Beyond Nature and Nurture*, edited by Susan McKinnon and Sydel Silverman, pp. 23-42. Chicago: University of Chicago Press.

Goodale, Jane C.
1980. Gender, Sexuality, and Marriage: A Kaulong Model of Nature and Culture. In *Nature, Culture and Gender*, edited by Carol MacCormack and Marilyn Strathern, pp. 119-142. Cambridge: Cambridge University Press.

Goodenough, Ward H.
1966 [1951]. *Property, Kin, and Community on Truk*. Hamden, CT: Archon Books.

Goodman, Alan, and Thomas L. Leatherman, editors
1998. *Building a New Biocultural Synthesis: Political-Economic Perspectives on Human Biology*. Ann Arbor: University of Michigan Press.

Gough, E. Kathleen
1968 [1959]. The Nayars and the Definition of Marriage. In *Marriage, Family, and Residence*, edited by Paul Bohannan and John Middleton, pp. 49-71. Garden City, NY: The Natural History Press.

Gould, Stephen Jay
1977. Biological Potentiality vs. Biological Determinism. In *Ever Since Darwin: Reflections in Natural History*, edited by Stephen Jay Gould, pp. 251-59. New York: W. W. Norton.

1980. Women's Brains. In *The Panda's Thumb: More Reflections in Natural History*, edited by Stephen Jay Gould, pp. 152-159. New York: W. W. Norton.

1981. *The Mismeasure of Man*. New York: W. W. Norton.

2000. More Things in Heaven and Earth. In *Alas, Poor Darwin: Arguments against Evolutionary Psychology*, edited by Hilary Rose and Steven Rose, pp. 101-126. New York: Harmony Books.

Grossberg, Michael
1985. *Governing the Hearth: Law and Family in Nineteenth-Century America*. Chapel Hill: University of North Carolina Press.

Harding, Sandra, editor
1983. *The "Racial" Economy of Science: Toward a Democratic Future*. Bloomington: Indiana University Press.

Harding, Sandra, and Jean F. O'Barr, editors
1975. *Sex and Scientific Inquiry*. Chicago: University of Chicago Press.

Helmreich, Stefan, and Heather Paxson
2005. Sex on the Brain: A Natural History of Rape and the Dubious Doctrines of Evolutionary Psychology. In *Why America's Top Pundits Are Wrong: Anthropologists Talk Back*, edited by Catherine Besteman and Hugh Gusterson, pp. 180-205. Berkeley: University of California Press.

Henninger, E.
1891. *Sitten und Gebräuche bei der Taufe und Namengebung in der Altfranzösischen Dichtung*. Thesis: Halle a. S.

Hill, Kim, and A. Magdalena Hurtado
1989. Hunter-gatherers of the New World. *American Scientist* 77:437-443.

1996. *Ache Life History: The Ecology and Demography of a Foraging People*. Hawthorne, NY: Aldine de Gruyter.

Hodgson, Geoffrey M.
1995. *Economics and Biology*. Brookfield, VT: Edward Elgar Publishing Company.

Howell, Signe
1989. *Society and Cosmos: Chewong of Peninsular Malaysia*. Chicago: University of Chicago Press.

Hua, Cai
2001. *A Society without Fathers or Husbands: The Na of China*. New York: Zone Books.

Hubbard, Ruth, and Elijah Wald
1999. *Exploding the Gene Myth*. Boston: Beacon Press.

Hutchinson, Sharon
1985. Changing Concepts of Incest among the Nuer. *American Ethnologist* 12:625-641.

Inhorn, Marcia C.
1996. *Infertility and Patriarchy: The Cultural Politics of Gender and Family Life in Egypt*. Philadelphia: University of Pennsylvania Press.

Inhorn, Marcia C., editor
2002. *Infertility around the Globe: New Thinking on Childlessness, Infertility, and the New Reproductive Technologies*. Berkeley: University of California Press.

Joyce, Rosemary A., and Susan D. Gillespie, editors
2000. *Beyond Kinship: Social and Material Reproduction in House Societies*. Philadelphia: University of Pennsylvania Press.

Kahn, Susan Martha
2000. *Reproducing Jews: A Cultural Account of Assisted Reproduction in Israel*. Durham: Duke University Press.

Kay, Herma Hill
1990. Perspectives on Sociobiology, Feminism, and the Law. In *Theoretical Perspectives on Sexual Differences*, edited by Deborah L. Rhode, pp. 74-85. New York: Yale University Press.

Kelly, Raymond C.
1976. Witchcraft and Sexual Relations: An Exploration in the Social and Semantic Implications of the Structure of Belief. In *Man and Woman in the New Guinea Highlands* (American Anthropological Association Special Publication 8), edited by Paula Brown and Georgeda Buchbinder, pp. 36-53. Washington, DC: American Anthropological Association.

Kirsch, A. Thomas
1975. Economy, Polity, and Religion in Thailand. In *Change and Persistence in Thai Society*, edited by G. William Skinner and A. Thomas Kirsch, pp. 172-196. Ithaca: Cornell University Press.

Koslowski, Peter, editor
1999. *Sociobiology and Bioeconomics: The Theory of Evolution in Biological and Economic Theory*. Berlin: Springer.

Krementsov, Nikolai L., and Daniel P. Todes
1991. On Metaphors, Animals, and Us. *Journal of Social Issues* 47(3):67-81.

Kummer, Bernhard
1931. Gevatter. *Handwörterbuch des Deutschen Aberglaubens*, Vol. 3. Berlin: Walter de Gruyter.

Leacock, Eleanor
1980. Social Behavior, Biology and the Double Standard. In *Sociobiology: Beyond Nature/Nurture? Reports, Definitions and Debate* (American Association for the Advancement of Science Selected Symposium 35), edited by George W. Barlow and James Silverberg, pp. 465-488. Boulder: Westview Press.

Lee, Richard B.
1965. *Subsistence Ecology of !Kung Bushmen*. Ph.D. Dissertation, University of California, Berkeley.

1979. *The !Kung San: Men, Women, and Work in a Foraging Society*. Cambridge: Cambridge University Press.

Lévi-Strauss, Claude
1969 [1949]. *The Elementary Structures of Kinship*. Boston: Beacon Press.

Lewontin, Richard C.
1991. *Biology as Ideology: The Doctrine of DNA*. New York, Harper Perennial.

Lewontin, Richard C., Steven Rose, and Leon J. Kamin
1984. *Not in Our Genes: Biology, Ideology, and Human Nature*. New York: Pantheon Books.

Lock, Margaret
1998. Perfecting Society: Reproductive Technologies, Genetic Testing, and the Planned Family in Japan. In *Pragmatic Women and Body Politics*, edited by Margaret Lock and Patricia A. Kaufert, pp. 206-239. Cambridge: Cambridge University Press.

MacFarquhar, Larissa
2001. The Bench Burner. *The New Yorker*, December 10:78-89.

Malinowski, Bronislaw
1929. *The Sexual Life of Savages in North-Western Melanesia*. New York: Halcyon House.

Marks, Jonathan
1995. *Human Biodiversity: Genes, Race, and History*. New York: Aldine de Gruyter.

Marshall, Lorna
1976. *The !Kung of Nyae Nyae*. Cambridge: Harvard University Press.

Martin, Emily
1991. The Egg and the Sperm: How Science Has Constructed a Romance based on Stereotypical Male-Female Roles. *Signs* 16(3):485-501.

McKinnon, Susan
1991. *From a Shattered Sun: Hierarchy, Gender, and Alliance in the Tanimbar Islands*. Madison: University of Wisconsin Press.

2000. Domestic Exceptions: Evans-Pritchard and the Creation of Nuer Patrilineality and Equality. *Cultural Anthropology* 15(1): 35-83.

2002. A obliteração da cultura e a naturalização da escolha nas confabulações da psicologia evolucionista. *Horizontes Antropológicos* [Brazil] 16:53-83.

2005. On Kinship and Marriage: A Critique of the Genetic and Gender Calculus of Evolutionary Psychology. In *Complexities: Beyond Nature and Nurture*, edited by Susan McKinnon and Sydel Silverman, pp. 106-131. Chicago: University of Chicago Press.

McKinnon, Susan and Sydel Silverman, editors
2005. *Complexities: Beyond Nature and Nurture*. Chicago: University of Chicago Press.

Mintz, Sidney W., and Eric R. Wolf
1968 [1950]. An Analysis of Ritual Co-parenthood (*Compadrazgo*). In *Marriage, Family, and Residence*, edited by Paul Bohannan and John Middleton, pp. 327-354. Garden City, NY: The Natural History Press.

Mintz, Steven, and Susan Kellogg
1988. *Domestic Revolutions: A Social History of American Family Life*. New York: Free Press.

Modell, Judith S.
1994. *Kinship with Strangers: Adoption and Interpretations of Kinship in American Culture*. Berkeley: University of California Press.

1998. Rights to Children: Foster Care and Social Reproduction in Hawai'i. In *Reproducing Reproduction: Kinship, Power, and Technological Innovation*, edited by Sarah Franklin and Helena Ragoné, pp. 156-172. Philadelphia: University of Pennsylvania Press.

Nelkin, Dorothy
2000. Less Selfish than Sacred?: Genes and the Religious Impulse in Evolutionary Psychology. In *Alas, Poor Darwin: Arguments against Evolutionary Psychology*, edited by Hilary Rose and Steven Rose, pp. 17-32. New York: Harmony Books.

Nelkin, Dorothy, and Susan M. Lindee
1995. *The DNA Mystique: The Gene as a Cultural Icon*. New York: W. H. Freeman and Company.

Orr, H. Allen
2003. Darwinian Story Telling. *The New York Review of Books* 50(3), February 27. http://www.nybooks.com/articles/16074.

Pinker, Steven
1997. *How the Mind Works*. New York: W. W. Norton.

2002. *The Blank Slate: The Modern Denial of Human Nature*. New York: Viking.

Posner, Richard A.
1981. *The Economics of Justice*. Cambridge: Harvard University Press.

1992. *Sex and Reason*. Cambridge: Harvard University Press.

Posner, Richard A., and Eric A. Posner
1998. The Demand for Cloning. In *Clones and Clones: Facts and Fantasies about Human Cloning*, edited by Martha C. Nussbaum and Cass R. Sunstein, pp. 233-261. New York: Norton.

Potts, Richard
  1996. *Humanity's Descent: The Consequences of Ecological Instability.*
  New York: William Morrow and Co.

  1998. Variability Selection in Hominid Evolution. *Evolutionary
  Anthropology* 7:81-96.

Powdermaker, Hortense
  1971 [1933]. *Life in Lesu: The Study of a Melanesian Society in New
  Ireland.* New York: W. W. Norton.

Ragoné, Helena
  1994. *Surrogate Motherhood: Conception in the Heart.* Boulder:
  Westview Press.

Rose, Hilary
  2000. Colonizing the Social Sciences? In *Alas, Poor Darwin:
  Arguments against Evolutionary Psychology*, edited by Hilary Rose
  and Steven Rose, pp. 127-154. New York: Harmony Books.

Rose, Hilary, and Steven Rose, editors
  2000. *Alas, Poor Darwin: Arguments against Evolutionary Psychology.*
  New York: Harmony Books.

Rose, Steven
  2000a. Escaping Evolutionary Psychology. In *Alas, Poor Darwin:
  Arguments against Evolutionary Psychology*, edited by Hilary Rose
  and Steven Rose, pp. 299-320. New York: Harmony Books.

  2000b. The New Just So Stories: Sexual Selection and the
  Fallacies of Evolutionary Psychology. *Times Literary Supplement*,
  July 14:3-4.

Sacks, Karen
  1974. Engels Revisited: Women, The Organization of
  Production, and Private Property. In *Women, Culture, and Society*,
  edited by Michelle Zimbalist Rosaldo and Louise Lamphere, pp.
  207-222. Stanford: Stanford University Press.

Sahlins, Marshall D.
  1976. *The Use and Abuse of Biology.* Ann Arbor: University of
  Michigan Press.

Sanday, Peggy Reeves
    1974. Female Status in the Public Domain. In *Women, Culture, and Society*, edited by Michelle Zimbalist Rosaldo and Louise Lamphere, pp. 189-206. Stanford: Stanford University Press.

    1981. *Female Power and Male Dominance: On the Origins of Sexual Inequality*. Cambridge: Cambridge University Press.

    1990. *Fraternity Gang Rape: Sex, Brotherhood, and Privilege on Campus*. New York: New York University Press.

Sapir, Edward
    1949 [1921]. *Language: An Introduction to the Study of Speech*. New York: Harcourt, Brace, and World.

Schiebinger, Londa
    1993. *Nature's Body: Gender in the Making of Modern Science*. Boston: Beacon Press.

Schlegel, Alice, editor
    1977. *Sexual Stratification: A Cross-Cultural View*. New York: Columbia University Press.

Schneider, David M.
    1980 [1968]. *American Kinship: A Cultural Account*. Chicago: University of Chicago Press.

Shostak, Marjorie
    1981. *Nisa: The Life and Words of a !Kung Woman*. New York: Vintage Books.

Singh, Devendra, Walter Meyer, Robert J. Zambarano, and David Farley Hurlbert
    1998. Frequency and Timing of Coital Orgasm in Women Desirous of Becoming Pregnant. *Archives of Sexual Behavior* 27(1)15-29.

Slocum, Sally
    1975. Woman the Gatherer: Male Bias in Anthropology. In *Toward an Anthropology of Women*, edited by Rayna R. Reiter, pp. 36-50. New York: Monthly Review Press.

Spencer, Robert F.
1968. Spouse-Exchange among the North Alaskan Eskimo. In *Marriage, Family, and Residence*, edited by Paul Bohannan and John Middleton, pp. 131-144. Garden City, NY: The Natural History Press.

Stocking, George W., Jr., editor
1974. *The Shaping of American Anthropology 1883-1911: A Franz Boas Reader*, edited with introduction by George W. Stocking, Jr. New York: Basic Books.

Strathern, Marilyn
1992. *Reproducing the Future: Anthropology, Kinship and the New Reproductive Technologies*. New York: Routledge.

Strathern, Marilyn, and Carol P. MacCormack, editors
1980. *Nature, Culture, and Gender*. Cambridge: Cambridge University Press.

Symons, Donald
1989. The Psychology of Human Mate Preferences (Commentary on Buss 1989). *Behavioral and Brain Sciences* 12:34-35.

Tanner, Nancy, and Adrienne Zihlman
1976. Women in Evolution, Part I: Innovation and Selection in Human Origins. *Signs* 1(3):585-608.

Tew, Mary
1951. A Form of Polyandry among the Lele of the Kasai. *Africa* 21(1):1-12.

Thomson, J. Anderson
2004. Why They Kill: Male Bonding + Religion = Disaster. *The Hook* 339, September 30: 26-29, 30.

Thornhill, Randy, and Craig T. Palmer
2000. *A Natural History of Rape: Biological Bases of Sexual Coercion*. Cambridge: MIT Press.

Todes, Daniel P.
1989. *Darwin without Malthus: The Struggle for Existence in Russian Evolutionary Thought*. New York: Oxford University Press.

Tooby, John, and Leda Cosmides
1989. The Innate Versus the Manifest: How Universal Does Universal Have to Be? (Commentary on Buss 1989). *Behavioral and Brain Sciences* 12:36-37.

1992. The Psychological Foundations of Culture. In *The Adapted Mind: Evolutionary Psychology and the Generation of Culture*, edited by Jerome H. Barkow, Leda Cosmides, and John Tooby, pp. 19-136. New York: Oxford University Press.

Travis, Carol
1992. *The Mismeasure of Woman*. New York: Simon and Schuster.

Trivers, Robert L.
1971. The Evolution of Reciprocal Altruism. *Quarterly Review of Biology* 46:35-57.

Tylor, Edward Burnett
1861. *Anahuac: or Mexico and the Mexicans, Ancient and Modern*. London: Longman, Green, Longman and Roberts.

Vayda, Andrew P.
1995. Failures of Explanation in Darwinian Ecological Anthropology: Parts I and II. *Philosophy of the Social Sciences* 25(2):219-49; 25(3):360-77.

Weiss, Rick
2005. Twin Data Highlights Genetic Changes. *The Washington Post*, Tuesday, July 5:A2.

Weston, Kath
1991. *Families We Choose: Lesbians, Gays, Kinship*. New York: Columbia University Press.

1995. Forever Is a Long Time: Romancing the Real in Gay Kinship Ideologies. In *Naturalizing Power: Essays in Feminist Cultural Analysis*, edited by Sylvia Yanagisako and Carol Delaney, pp. 87-110. New York: Routledge.

Wheeler, David
1996. Evolutionary Economics. *The Chronicle of Higher Education*. July 5: A8, A12.

Whyte, M. K.
1978. Cross-Cultural Codes Dealing with the Relative Status of Women. *Ethnology* 17:211-237.

Wilson, Margo, and Martin Daly
1992. The Man Who Mistook His Wife for a Chattel. In *The Adapted Mind: Evolutionary Psychology and the Generation of Culture*, edited by Jerome H. Barkow, Leda Cosmides, and John Tooby, pp. 289-326. New York: Oxford University Press.

Wright, Robert
1994. *The Moral Animal: The New Science of Evolutionary Psychology*. New York: Vintage Books.

Yanagisako, Sylvia, and Carol Delaney, editors
1995. *Naturalizing Power: Essays in Feminist Cultural Analysis*. New York: Routledge.

Zihlman, Adrienne L.
1978. Women in Evolution, Part II: Subsistence and Social Organization among Early Hominids. *Signs* 4(1):4-20.

Also available from Prickly Paradigm Press: